T0134978

Studies in Systems, Decision and Control

Volume 143

Series editor

Janusz Kacprzyk, Polish Academy of Sciences, Warsaw, Poland
e-mail: kacprzyk@ibspan.waw.pl

The series "Studies in Systems, Decision and Control" (SSDC) covers both new developments and advances, as well as the state of the art, in the various areas of broadly perceived systems, decision making and control- quickly, up to date and with a high quality. The intent is to cover the theory, applications, and perspectives on the state of the art and future developments relevant to systems, decision making, control, complex processes and related areas, as embedded in the fields of engineering, computer science, physics, economics, social and life sciences, as well as the paradigms and methodologies behind them. The series contains monographs, textbooks, lecture notes and edited volumes in systems, decision making and control spanning the areas of Cyber-Physical Systems, Autonomous Systems, Sensor Networks, Control Systems, Energy Systems, Automotive Systems, Biological Systems, Vehicular Networking and Connected Vehicles, Aerospace Systems, Automation, Manufacturing, Smart Grids, Nonlinear Systems, Power Systems, Robotics, Social Systems, Economic Systems and other. Of particular value to both the contributors and the readership are the short publication timeframe and the world-wide distribution and exposure which enable both a wide and rapid dissemination of research output.

More information about this series at http://www.springer.com/series/13304

Mauricio A. Sanchez · Leocundo Aguilar
Manuel Castañón-Puga · Antonio Rodríguez-Díaz
Editors

Computer Science and Engineering—Theory and Applications

 Springer

Editors
Mauricio A. Sanchez
Calzada Universidad
Tijuana, Baja California
Mexico

Manuel Castañón-Puga
Calzada Universidad
Tijuana, Baja California
Mexico

Leocundo Aguilar
Calzada Universidad
Tijuana, Baja California
Mexico

Antonio Rodríguez-Díaz
Calzada Universidad
Tijuana, Baja California
Mexico

ISSN 2198-4182   ISSN 2198-4190   (electronic)
Studies in Systems, Decision and Control
ISBN 978-3-319-89267-2   ISBN 978-3-319-74060-7   (eBook)
https://doi.org/10.1007/978-3-319-74060-7

Preface

The fields of computer science and computer engineering are rich within themselves such that they are comprised of a vast amount of areas, such that most books decide to better separate their contents into specialized areas, ignoring the rest of their field. But this book takes on a different point of view when compared to most of existing literature, that is, to collect from different areas within these two fields to demonstrate the great variety which exists, where by harboring different contributed research chapters in a single book, instead of granular specialized areas, the concept of the complete fields is once again grasped.

As computer science and computer engineering are two different fields, they have more in common than differences; therefore, the objective of this book is to convey their involvement in society by showing advances in theoretical practices, new uses for existing concepts, and applications.

This book is intended as a reference for pre-graduate students who wish to know more about advances in the fields of computer science and computer engineering, or acquire ideas as to what types of areas can be researched, exploited, or applied; for graduate students and researchers, who might need to know the current state of included contributed research, as well as acquire ideas for their own research; and for professionals, who would want to know the current state of their fields, get ideas to solve problems at their workplace, or simply to get a sense of direction as to where these fields are going.

The contents of this book traverse various areas in the fields of computer science and computer engineering, such as software engineering, computational intelligence, artificial intelligence, complex systems, systems, engineering, and embedded systems.

Provided diversity in the contents of this book is the strength between its covers, we hope that readers enjoy our book and find some utility within its covers. And that they start seeing that the fields of computer science and computer engineering can also be seen as complete and diverse fields, instead of only parts of them.

Tijuana, Mexico   Mauricio A. Sanchez
November 2017   Leocundo Aguilar
                                                                        Manuel Castañón-Puga
                                                                       Antonio Rodríguez-Díaz

Contents

A Comprehensive Context-Aware Recommender System Framework

Sergio Inzunza and Reyes Juárez-Ramírez

Abstract Context-Aware Recommender System research has realized that effective recommendations go beyond recommendation accuracy, thus research has paid more attention to human and context factors, as an opportunity to increase user satisfaction. Despite the strong tie between recommendation algorithms and the human and context data that feed them, both elements have been treated as separated research problems. This document introduces MoRe, a comprehensive software framework to build context-aware recommender systems. MoRe provides developers a set of state of the art recommendation algorithms for contextual and traditional recommendations covering the main recommendation techniques existing in the literature. MoRe also provides developers a generic data model structure that supports an extensive range of human, context and items factors that is designed and implemented following the object-oriented paradigm. MoRe saves developers the tasks of implementing recommendation algorithms, and creating a structure to support the information the system will require, proving concrete functionality, and at the same time is generic enough to allow developers adapt its features to fit specific project needs.

1 Introduction

Recommender systems or recommendation systems are information filtering tools that alleviate information overload to users, mainly by predicting the preference of the user for certain items and presenting the users the items more likely to be consumed by the user [1].

Context-Aware Recommender Systems (CARS) aims to further improve recommendation accuracy and user satisfaction by taking into account contextual information [2]. The inclusion of this new information into the recommendation process has proved to help increasing prediction accuracy of recommender systems [3].

S. Inzunza (✉) · R. Juárez-Ramírez
Universidad Autónoma de Baja California, Tijuana, Baja California, Mexico
e-mail: sinzunza@uabc.edu.mx

© Springer International Publishing AG, part of Springer Nature 2018
M. A. Sanchez et al. (eds.), *Computer Science and Engineering—Theory and Applications*, Studies in Systems, Decision and Control 143,
https://doi.org/10.1007/978-3-319-74060-7_1

1

CARS are based on the idea that similar user in similar context will like similar items, and that the user preferences for certain item change according to the contextual situation [4].

In the beginning, the research in CARS is leaned toward the development of new, and optimizing the existing algorithms to support contextual information and to generate better recommendation by improving the accuracy [5].

In recent years, researchers have become more aware of the fact that effectiveness of recommender systems goes beyond recommendation accuracy. Thus, research in the context and human factors has gained increased interest [6] as a potential opportunity to increase the user satisfaction with the recommendation results.

Despite the strong tie between the recommendation algorithms and the data about user and context factors, both elements have been treated as separated research problems. Because of this separation, to implement a CARS nowadays, software developers, and architects must divide their efforts into 2 topics: (i) The recommendation system topic to choose and implement the best algorithm for their CARS domain; (ii) The second topic that also requires a significant amount of effort is the user modeling. In the first topic, according to [7], the implementation of recommendation algorithms can be a complicated task, especially for developers who are not experts in the field [5]. As for the second topic, software architects need to come up with a data model that is capable of holding all the needed information for the algorithms to work, if the model designer has no experience with CARS over what aspects to model for better recommendations, this can result in an incomplete and overspecialized model [8], that can limit the functionality of the recommendation algorithms.

A software framework to support the development of advanced, complex contextual recommender systems should provide the developer both main elements of a CARS, implemented algorithms from different recommendation techniques so developer can select the one that better work for CARS item domains (e.g. songs, movies, etc.) [9], as well a data model structure generic enough to work with most CARS domains and give developers the option to adapt it to suit the specific need of the project. Such a framework is, to the best of our knowledge, currently no available. Even when some framework and libraries proposals exist to help in the creation of contextual recommender systems, they are focused on either modeling the user and context information [10] and do not include recommendation features, or focus only in the algorithms for recommendation (like Hybreed [5]) and don't considered modeling the user and context information and rely on the information being in a dataset file, which is not practical for a real-world applications.

In this paper, we introduce *MoRe (Modeling and Recommendation framework)*, a software framework designed to support the creation of CARS by proving developers a set of state of the art contextual algorithms, and a generic user model capable of structure and manage the information required by the CARS. We consider MoRe to be a *comprehensive* framework, as in a single proposal, provides means to solve both main problems developers encounter when implementing CARS (i.e. data modeling and recommendation algorithms). MoRe provide developers object-oriented classes and methods they can use to store, retrieve and perform recommendation over the user, context and item data.

MoRe aims at serving as a tool for new CARS development, helping developers in the algorithm implementation by providing ready to work algorithms from main recommendation techniques for multidimensional data (transformation and adaptation). MoRe also helps in the creation of the data architecture by proving an extensive class architecture capable of organizing, persisting and retrieving data about the user, context, and items for most popular recommendation domains, like movie, song, hotels, restaurants, and travel recommendation among others. For the data structure, uses GUMORS, a General User Model for Recommender System presented in our previous work [11].

The rest of this document is organized as follows: Sect. 2 describes the background and related work, Sect. 3 describes the MoRe framework, Sect. 4 present some evaluation performed, and Sect. 5 present conclusion and future work.

2 Background and Related Work

The implementation of CARS requires modeling of the data and implementation of algorithms that use such data to generate predictions, as mentioned earlier, both tasks are treated separately in literature. The same way, in this section we review and analyze the most relevant proposals related to modeling the context for adaptive systems and compare them with [11], the data model used in MoRe. Then we review and analyze most relevant proposals of software libraries and frameworks created to generate context-based recommendations.

This section concludes with a set of requirements derived from literature that a software framework intended to help developers in the creation of CARS should fulfill.

2.1 Context and Context Awareness

The definitions for context varies depending in the domains of applications, in computer science Schilit et al. [12] described context as a union of three aspects: "the location of user, the collection of nearby people and objects, as well as the change to those objects over time", but such definition is rather broad. Dey and Abowd [13] provide another definition that is commonly accepted in computer science areas:

> Context is any information that can be used to characterize the situation of an entity. An entity is a person, place or object that is considered relevant to the interaction between a user and an application, including the user and the applications themselves.

Certainly, based on this definition of context, the user plays an important role, the context of a user can include any information that describes his situation, like his location and time, emotional, mental and physiological information, etc. The amount and type of information to include in the contextualization of the user depends on the system and the purpose of such information.

A system is considered to be context-aware if it can express aspects of the environment and subsequently uses such information to adapt its functionality to the context of use [14]. Therefore, context-awareness refers to the capability of an application being aware of its physical environment or situation and responding proactively and intelligently based on such awareness.

Context-aware systems aim at somehow gathering (through sensing, inferring or directly asking for) human and environmental phenomena for later assisting users to archive a desirable quality of living standards [15].

From an *informational* perspective, context provides information that systems can use to form a contextual space, which virtually represents the situation of the user [16].

From a software *infrastructural* perspective, context provides computing devices with information about its environment as provided by other parts or modules of the system (sensing module for example). As a consequence, different 'types' or 'dimensions' of context emerge, e.g. physical and computation dimensions.

2.1.1 Context Representation

There are several techniques that will allow a developer to represent the contextual information inside a computational system, thus to be used in a CARS. Works like [17–19] present extensive surveys on the difference of each technique. Next, we describe and discuss the most commonly used representation techniques at high-level, and present the main advantages and disadvantages of each one.

- *Key-Value models.* These models use pairs of a key and value to enumerate attributes (key) and their values to describe the contextual and user information. These models are the simplest data structure and are easy to manage, especially when they have a small amount of data. However, key-value modeling is not scalable and not suitable to represent complex data models [19]. This technique is best suited to represent and store temporary information, therefore is increasingly less used in recent contextual and user models.
- *Markup scheme models.* These models use a hierarchical data structure formed by markup tags with attributes and content. To represent the user and context aspect, markup models use a set of symbols and annotations inserted in a text document that controls the structure, formatting, and relations among annotations [20]. As markup languages do not provide advanced expressive capabilities, reasoning over the data they represent is hard. Further, retrieval, interoperability, and re-usability of the data over different models can be difficult, specifically if such models use different markup schemes [17].

- *Ontology-based models.* Ontologies represent a description of the concepts and relationships. Ontologies incorporate semantics into XML-based representation or Resource Description Framework [21]. Ontology-based context models are fairly common, because of the formal expressiveness and the ability of ontologies to support reasoning. However, is hard to construct complete ontologies and avoid the ambiguity in the ontology [18]. Also the information retrieval can be computationally intensive and time-consuming when the amount of data is increased [17].
- *Graphical models.* These models are mainly based on the Unified Modeling Language (UML) [22] and Entity-Relationship Model (ERM) [23]. UML is a standardized general-purpose modeling language used in software architecture description, which can represent the user and context aspects, as well as its relations. Graphical models are capable of expressing the modeled aspects by graphical diagrams, and regarding expressive richness, graphical modeling is better than markup and key-value modeling as it allows relationships to be captured into the model [17].
- *Object-oriented context model.* These technique models the data by using object-oriented techniques, this offers the full power of object orientation like encapsulation, reusability, and inheritance. In most cases, this model technique encapsulates the processing of context at the object level and allows instances to access context and user information by inheritance mechanism [18]. The main flaw of this technique is the lack of direct model validation, and when complex models are created, it may not be supported by limited resources hand-held devices [19]. Nevertheless, as most of the high-level programming languages support object-oriented concepts, these models can be integrated into context-aware systems easily. This makes object-oriented modeling to be used as code base, run-time, manipulation, and storage mechanism for user and context models.

As there are various model representation techniques, each with their advantages and disadvantages, a challenged to overcome in our development was to choose the one that better suit the need to represent a data model for a context-aware recommendation systems framework. We opted to create a combination of graphical modeling, which allowed us to create a reach and expressive data model that contains the intrinsic relationships of the user, context and items information.

2.2 Frameworks for Context-Aware Recommendations

Context-Aware Recommendation Systems has gained a lot of attention in the personalization and data mining communities, as a result, there exist a lot of literature focused on different aspects of CARS. Next, we review and analyze CARS literature that proposed frameworks aimed at supporting developers in the implementation of these type of systems and compares how each of the presented proposals relates to MoRe.

MyMediaLite [24] is a recommendation library implemented in C# and aimed towards recommending items based on collaborative filtering recommendation technique. Even when MyMediaLite was designed to work in traditional (2D) recommendations, it has been used by [25] as a base-line recommender in context-aware recommendation. MyMediaLite is similar to our proposal in the fact that both provide the feature of 2D collaborative-filtering algorithms, but MoRe goes far beyond providing nD recommendation algorithms.

Hybreed [5] is a Java-based recommendation framework designed specifically to help developers in the creation CARS. Hybreed focuses on the integration of external context to recommendation systems and the hybrid combination of different basic recommendation methods. A notable feature of Hybreed is its dynamic contextualization approach that creates an in-memory temporal model of the user containing his current situation. Hybreed and MoRe had in common the feature of proving developers working ready algorithms to generate contextual prediction based on filtering techniques. But unlike More, Hybreed do not provide recommendation algorithms based on the context-modeling technique. In addition, Hybreed stands behind in the data modeling technique, as it uses a limited Key-Value pair, while MoRe uses a context-aware user model.

ConRec [1] is a context-aware recommendation framework focus on serving as a tool for developers when implementing CARS, ConRec mainly focuses on a temporal dynamic representation of context, which can automatically aggregate different contextual features into a single one, so it can be easier for algorithms to process. ConRec includes its own multi-dimensional recommendation algorithms that works as an extension to Apache Mahout. This proposal does not describe how user or context factors are considered into the framework, nor how they are stored. Compared to MoRe, ConRec includes only one algorithm, that is not supported by other literature, while MoRe contains implementation for various algorithms that are well supported by previous research. Also, MoRe present an clear and well-structured manner of managing the data, while ConRec only mentions that uses it a *user, item, rate* and *context* format, and no information in how to take such approach to implementation is described.

CoGrec [26] is a theoretical recommendation framework designed to gather latent features of a group of users based on their social networks then uses such gathered information to predict items that better suit the preferences of all the individuals in the group. This proposal is described theoretical, and even when some evaluation results are presented, no implementations details were given. CoGrec differs from MoRe, on being a group-based recommendation only, while MoRe is flexible in the target user(s). Also, MoRe is closer to implementation providing a working-ready framework.

In [27] a context-aware architecture for mobile recommendation systems are proposed. The architecture is designed to be generic enough to work in any CARS domain, and focus on supporting the mobility of CARS, and the communication between the mobile device and the server where the recommendation take place.

A downfall of this architecture only provides a template for pre-filter recommendations only, while MoRe provides templates and implementation for a range of recommendation techniques. MoRe is also ahead of this proposal in the management and modeling of the data, as the proposal only provides the means of storing and retrieving the information, while MoRe provides, this functionality along with the design of a generic data structure.

CARSKit [28] is a Java-based algorithms library, which contains implementations for various recommendation techniques, for traditional (e.g. average and collaborative filtering) and contextual (e.g. item splitting and context-aware matrix factorization) recommendations. Even when CARSKit contains one of the most complete set of algorithms, these are designed for scientific purposes, focused mainly in the evaluation of algorithm results, and are not capable of recommending items to a specific user, which is the main function of the algorithms in a real CARS implementation. MoRe uses some of the algorithms implementation of CARSKit and adapted them to be able to recommend items to users. CARSKit and More differ in their nature, as CARSKit is a set of algorithms, while MoRe is a recommendation framework that as part of its features, contains a set of algorithms.

2.3 Requirements for a Context-Aware Recommender System Framework

Based on related literature presented above, an in the work of [5, 15, 29, 30], this section describes a series of high-level functional requirements any software framework aimed to facilitate the creation of context-aware recommender systems should fulfill.

First of all, such a framework should include recommendation algorithms from the most relevant contextual recommendation techniques, which developers can set up with a few lines of code, and be able to obtain the list of recommended items. These algorithms should cover the main techniques existing nowadays, which according to Adomavicius [31] are Contextual Filtering and Contextual Modeling. To support contextual modeling recommendation, the framework should include multi-dimensional recommendation algorithms, and to support contextual filtering, traditional recommendation algorithms are needed as well. Apart from the implemented and ready to work algorithms, the framework should support the integration of custom techniques and new algorithms.

With regard to the data needed by CARS, the modeling of contextual information has been treated separately from the recommendation functions, and none of the currently existing frameworks support the management of the CARS data in the same proposal as contextual recommendation algorithms. Therefore, a comprehensive contextual recommendation framework should provide a baseline data structure that supports the data about the users of the system, their context, the items that the CARS will recommend, and the relations among the different aspects. In this aspect, diversity has a great importance [6], as the framework should support CARS from different items domains, i.e. the framework should be able to recommend movies, as well as restaurants or pets.

Besides this specific requirements, a framework targeting developers should meet general software engineering requirements, for example, the set of guidelines for software architecture proposed by Microsoft Application Architecture Guide [32]. Such a framework should make easy for developers to adapt existing and to add new functionality.

According to [5], the expected solution is not as simple as combine all features from different proposal that meets certain requirements to create a framework that covers them all. The challenge is to select the most valuable concepts of all these existing approaches, reduce their programming complexity and combine them, such that the resulting framework coverts the described requirements with a rich set of functionalities, and at the same time is easy to use and reduce development effort.

3   The Comprehensive Context-Aware Recommender System Framework

This section describes the MoRe (**Mo**deling and **Re**commendation) framework that was created to support the creation of context-aware recommender systems, by providing a comprehensive class architecture that can model and manage the information required for the CARS systems to work, and by incorporating a set of state of the art recommender algorithms.

MoRe target, but is not limited to, the following users:

- *Software developers* who want to create a new CARS, either with or without experienced creating this type of applications. To these users MoRe offers a data modeling feature that developers can rely on to manage the information needed by the recommender algorithms. MoRe also includes a set of state of the art that with a few lines of code, developers can set up to generate the contextual recommendations.
- *Software architects* who want to create a data architecture to support the user, context and item information for a CARS. To these users MoRe presents an extensible class structure that supports all this information, and they can extend or user it straight into the system architecture.
- *Researchers* of recommendation systems area, who want to compare existing recommendation algorithms with a new approach, as they can use load their dataset(s) into MoRe data model and perform recommendations over it using the contextual algorithms included. Or researchers who just want to put their algorithms to test in a real scenario, and don't want to spend time designing a data model.

Fig. 1 Architectural view of
MoRe framework

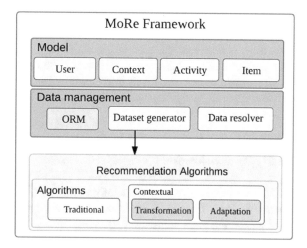

MoRe is implemented as a C# Framework following the Object-Oriented Paradigm (OOP). MoRe can be set up with a few lines of code, and can help in the modeling and management (storage and retrieval) of user, context, and item information, as well in the process of generating contextual recommendations based on the acquired information. The central element of the framework is a context-aware user model designed specifically for CARS, that along with a data management module, and with a recommendation algorithms constitute the proposed MoRe framework.

Next, Fig. 1 presents an overall view of MoRe framework architecture, then each component of the framework is described. Section 3.1 describes the data model, 3.2 describes de data management, and Sect. 3.3 describes the algorithms included in MoRe framework.

3.1 The Model

In our previous work [11] a Generic User Model for context-aware Recommender System (GUMORS) was proposed to solve the problem of not existing an extensive model that can be used as a reference to structure the user and context aspects inside CARS. GUMORS provides a large set of user, context, and items working together to create a CARS specific data model, that can be used into multiple CARS domains. MoRe uses GUMORS as data structure to manage all the information needed by prediction functions.

GUMORS organize the CARS information into 4 main top-level categories, namely: *User*, *Context*, *Item*, and *Activity* information. Next, the four top-level categories are briefly described, for a more detailed description please refer to [11].

3.1.1 User Aspects

User aspects represent the user as a human being, which CARS can use to infer the user preferences and behavior to better suit the recommendation results.

User aspects are categorized into Contact, Demographic, Physiological, Interest and Preferences, Role, Personality, Emotion, and Mental information. Next Table 1 describes each category.

3.1.2 Context Aspects

GUMORS represent the context information in 6 categories, namely: *Computing, Location, Time, Physical Conditions, Resource* and *Social Relation.* Next Table 2 briefly describe each context category.

3.1.3 Activity Information

The activities information that can be useful for CARS are also modeled into GUMORS. An activity relates information about the *User* that is performing it, in a specified *Context.* A more specialized type of activity is called *RatedActivity* which also includes information about the item consumed during the activity, and the feedback (*Rating*) the user provided. The activity information is used for example by Google Maps, to suggest the user navigation route according to his displacement activity (walking, biking, driving, etc.). Spotify also uses activity information to

Table 1 User aspect considered by GUMORS

Category	Category description
Contact	Refers to identify the user from the rest of the user in the system, this also includes information about the user account in the system like the full name of the user, his email and address
Demographic	Describes the demographical information of the user like gender, birthday and language that the user
Emotion	Represents information about the subjective human emotions of the users
Interest and preferences	Explicitly describes the interests and preferences of the user for certain items, and can also be used to store system preferences (e.g. font size)
Mental	Used to describe the user's state of mind, this category includes information about user's mood; mental state and cognitive style that can be used to provide more tailored recommendations
Personality	Describes permanent or very slow changing patterns that are associated with an individual
Physiological	Models the aspects of the human body and its functionality
Role	This category represents the roles the users play during their activities

Table 2 Contextual information considered by GUMORS

Category	Description
Computing	This category refers to information about computational elements that the user is interacting with, or that are blended into the environment
Location	Refers to information that relates an item, user, or other context information with a geographical position. Location include information about the *Place* either *Physical* or *Digital*, *Address* and *Coordinates* for physical
Time	Used to capture information about time, such as Date, *TimeOfYear*, *TimeOfDay*, etc.
Physical condition	Describes environmental conditions where the system or user is situated at certain point on time
Resource	Model relevant characteristics of the physical or virtual environment
Social relations	Refers to social associations or connections between the user of the system and other persons

Table 3 Classes used by GUMORS to represent the activity information category

Class	Description
Activity	Main class used to represent the activities. This class has attributes like the *Time Stamp*, *Name* of the activity, and relations to a *User* and a *Context*
Rated activity	This class is a specialization of the *Activity* class. This class includes information of the *Item* involved in the activity, and the *Rating* the user gives to such experience. GUMORS include the following concrete specialization of *Rated Activity: View, Listen, Eat, Travel* and *Purchase*

recommend users songs which rhythm matches his running speed. Next Table 3 shows the classes used by GUMORS to model the activity information.

3.1.4 Items Information

GUMORS also manages the information about the Items the CARS will recommend. For this feature, GUMORS uses an *Item* super-class that developers can further specialize through inheritance to fit their specific needs. GUMORS also contains a series of *Item* specialization based on the most commonly used items in CARS literature and considers the commonly used attributes for each item (as shown in Fig. 2).

3.2 Data Management

The data management module is in charge of persisting the data from the model to a database and retrieving it back when asked. MoRe uses Entity Framework [33] for the Object-Relational Mapping (ORM). The use of an ORM for the data persistence

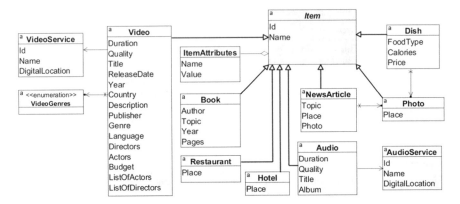

Fig. 2 Classes used by GUMORS to model the items of CARS

allows the framework to perform CRUD (Create, Read, Update and Delete) operations directly over the model classes that is automatically translated into the database without the need to directly perform queries to the db.

MoRe uses a *ModelEntity* superclass that contains generic methods with the logic for all the CRUD operations, using both, synchronous and asynchronous calls. This superclass is inherited from all the model classes that should be mapped to database, which inherits the CRUD operations, e.g. *Find, FindAll, Save, Delete, Get, GetAll*, etc.

3.2.1   Dataset Generator

The dataset generator is responsible for collecting the data from the database and organize it so recommendations algorithms can understand it, e.g. Comma Separated Values (CSV) or Binarized format [34].

MoRe uses *Data Annotations* to annotate the classes and attributes that will be part of the resulting dataset. Annotating classes and attributes work as rules that latter the framework will use to automatically create the dataset, which can be read by algorithms from memory. As MoRe is designed to work custom internal or external algorithms, the dataset can also be exported to a file in the specified format, so any existing algorithm can use the data to generate recommendations.

3.3   Recommendation Algorithms

MoRe uses a large set of algorithms for both, traditional (2D) and contextual (nD) recommendation. For this first iteration, MoRe uses the set of algorithms implemented by [28] and a custom wrapper to execute them from C# as originally the implementation were made under Java development language.

As the framework contained a large set of ready to use algorithms, it can be used by developers who don't want to spend too much time implementing existing recommendation techniques, or by developers that have little experience implementing this type of complex systems [5], the recommendation algorithms can be used as a black-box, that need to fed whit a dataset (described in previous section), and will yield a list of recommended item for the specified user.

The framework also allows for custom algorithms to be included, which makes MoRe ideal for researchers who want to test their new algorithm or recommendation approach against existing state of the art algorithms.

Next, the traditional and contextual recommendation algorithms are described.

3.3.1 Traditional Recommendation Algorithms

Even when MoRe is created to be a context-aware recommendation framework, it includes traditional recommendation as they are used by filtering recommendation techniques, described in later section, and even when is not the goal, the framework can be used for traditional recommendation systems. For this iteration, MoRe include some of the main traditional recommendation algorithms for every category (as shown in Table 4), based on the implementation provided by LibRec [35].

3.3.2 Contextual Recommendation Algorithms

The multi-dimensional recommendation algorithms are the core of any recommendation framework, therefore MoRe provides several different algorithms from state of the art techniques, which gives developers the ability to choose the algorithm that yields better results for their specific domain, as some algorithms may perform better when recommending movies based on user's companions, and others perform better when recommending songs or books.

Table 4 Traditional recommendation techniques and algorithms supported by MoRe

Technique	Brief description	Algorithms
Ranking recommendation	This recommendation technique build a recommendation model by analyzing the importance of each with respect all other items in the system [36]	*SLIM BPR*
Average recommendation	A simple recommendation technique that perform prediction by the average rating given to an element [37]	*GlobalAvg* *UserAvg* *ItemAvg*
Collaborative filtering recommendation	Collaborative-Filtering (CF) techniques construct a model of similarities between users or items based on the idea that if two users gave a similar rating for a specific movies, the users or items are very similar [38]	*UserKNN* *ItemKNN* *SVD ++*

In addition to the list of implemented and ready to work algorithms, MoRe allow developers to implement their own algorithms o to create hybridizations combining multiple of the included ones. To implement new algorithms, developers need to implement the *IRecommender* or *IContextualRecommender* class for a new traditional or contextual algorithms respectively and implement the methods each interface enforces.

Contextual Filtering Algorithms, try to pre- or post-process the information and convert it from a multi-dimensional matrix to a 2-dimensional rating matrix that contains only the user reference (Id), item reference, and rating. Then, this two-dimensional matrix can be used in traditional algorithms to generate recommendations. There exist many contextual filtering algorithms, one of the most effective is the context-aware splitting [39]. MoRe includes the three main variants of this algorithm:

– *UserSplitting*: From a user point of view, the preference for certain items may vary depending on the context. User Splitting group items based on the context and the rating the user gave to them, so when a user gives items good values in a certain context (e.g. sunny weather) and gives bad rating in another context situation (e.g. rainy weather), the user can be divided into two users based on such context (weather in this example).
– *ItemSplitting*: Separate the item that was rated differently under different context as being different items. Based on the same rules of *UserSplitting* to decide when some instances should be divided, but this process divide items instead of users.
– *UserItemSplitting*: Combines both previous separations, the result is that an item rated by a user in different context, is converted into 2 items, and the user is also divided into 2 users.

Contextual Modeling Algorithms, unlike splitting approaches, take into account the influence of context all context aspect on the rating prediction model, which required the development of new algorithms capable of processing multi-dimensional matrix and produce a recommendation. The algorithms used for contextual modeling supported by MoRe are:

– Tensor Factorization (TF) [40] This algorithm is based on the Matrix Factorization (MF) dimensional reduction technique which is used for 2D recommendations. TF consist of extending the two-dimensional MF problem into a multi-dimensional version, where the rating tensor is converted into a lower-dimensional vector space. Such that the interactions between users, items, and contextual factors are represented by a latent factor vector.
– Context-Aware Matrix Factorization (CAMF) [41] is a more scalable (than TF) contextual modeling approach based on MF. CAMF uses a contextual baseline prediction function to represent the interactions of contextual information with the items or users. Baltrunas et al. [42] proposed different variants of CAMF that model the influence of contextual conditions at different granularities. *CAMF-C* assumes that a context factor affects the user and items in the same way.

CAMF-CI models the in influence of a contextual aspect over items. And *CAMF-CC* assumes the context affects the ratings for all the items, ignoring the effect of context over users. MoRe contains the three variants (*CAMF-C, CAMF-CI* and *CAMF-CC*) variants of this algorithm.

Custom Context-based Algorithms, MoRe supports the addition of new algorithms defined by developers and researchers, which allow them to use the framework to implement and put their theoretical algorithm to test. Any new contextual algorithm that will be added to MoRe must implement the *IContextAwareRecommender* interface, which defines a method to build, save and load the prediction model, as well as the recommend method what should return a list of *RecommendedItem*s.

4 Evaluations

Assessing the quality of complex software systems is generally a difficult task, especially when the software provides new functionality that is not supported by other software. As previously mentioned, MoRe is the only one of its kind, comprehensive Context-Aware Recommender Systems framework. Therefore a direct comparison with any other proposal can yield inaccurate and misleading results.

To validate MoRe applicability and effectiveness, first, we perform comparative assessments with other frameworks and libraries' features. Then some use cases are implemented using MoRe. For both tests, Microsoft Visual Studio 2017 was used as development IDE, along with Sql Server 2012 express as database engine, with a Core i7, 16 GB Ram on Windows 10 computer.

4.1 Comparative Assessment

This section compares MoRe with other framework and libraries discussed in Sect. 2 of this document.

Next Table 5 presents a comparison between MoRe and other related context-aware recommender frameworks. The *Type* section describes whether the proposal is presented as a theory, a library or a framework; the *Data* section describes if the proposal presents a way to structure the information about the user, context, and items, and if it supports a temporal state profile which maintains a sub-profile of the user with the current information so it can be quick and easily accessible by the application. The *Algorithm* section describes the recommendation techniques supported by each proposal, and finally, the last section describes the programming language used by the proposal and if it is available as an open source (OS), not disclosed (ND) or not available (NA).

Table 5 Comparison of recommendation frameworks

	MoRe	Hybreed	MyMediaLite	ConRec	CoGrec	CARSKit
Type						
Theoretical					•	
Library			•	•		•
Framework	•	•				
Data						
User model	•	○		○	○	
Context model	•			○	○	
Item model	•					
Temporal state		•		•	○	
Algorithms						
Traditional						
Ranking	•		•			○
Average	•	•	•			○
Col. filtering	•	•	•			○
Contextual						
Contextual filtering	•		○			○
Contextual modeling	•	•		•	•	○
Group recommender		•			•	
Other						
Language	C#	Java	C#	Java		Java
Availability	OS	OS	OS	ND	NA	OS

• Denotes full support; ○ Partial support; OS: Open Source, ND: Non Disclosed

MoRe is compared against Hybreed [5], MyMediaLite [24], ConRec [1], CoGrec [26] and CARSKit [28].

4.2 Use Cases

MoRe is framework created to support the creation of context-aware recommender systems, so is very important the framework work in real-world usages as described conceptually. To test the applicability of the framework in CARS developments, and to validate that the resulting framework fulfills the requirements described in Sect. 2, some use cases where implemented. These use cases represent essential task of any CARS development (i.e. data storage and retrieval, and performing recommendations on the stored data).

Fist, the data management feature of the framework is tested selecting a real-world dataset from literature and loaded into MoRe's data model. Then, the ability of the framework to generate recommendations is tested using the loaded data, and generating predictions using different recommendation techniques.

4.2.1 Data Management Feature

Part of the MoRe's core is a data structure capable of modeling the users, context, and items information that will be used by recommendation algorithm to generate predictions. To test this feature, the LDOS CoMoDa [43] dataset was used loaded into MoRe's model. LDOS CoMoDa is a movie dataset that contains 2296 instances, and 32 attributes. The attributes are 9 corresponding to user information, 9 of contextual information and 14 attributes used to describe the items (movies).

Process

The process followed was to set up a new C# project in Visual Studio, and add references to MoRe framework. As the dataset's data is separated in 3 files, first the *Items* information was loaded into the model by reading the *itemsTitles* and *itemsMetaData* files, then the user, context and ratings information was loaded from the *LDOS-COMODA* file. Next Table 6 show the user, context and items aspects considered in the dataset and how they were mapped to MoRe data model.

Once the data was loaded into the model, and to test the ability of the framework to serve the data, a dataset was creating using class annotation (as described in Sect. 3.2) to configure the exportation behavior of the classes. Such recreated dataset set was used in the next section of use case, where the functionality of the recommendation algorithms is tested.

Results

The use case of loading data to MoRe's model showed that the framework was capable of supporting all LDOS CoMoDa dataset features, most of them (93%) without the need of any adaptation to the model. Two features (Decision and Interaction, the first refers to what motivate the user to watch the video, and the latter refers to the number of interactions of the user with the video) were not directly supported by the model as shipped with the framework.

To support these two features into the model, some adaptations were required, the adaptations consisted of further specializing (through inheritance) the *Video* and *ModelORM* classes, adding the required features to the new video (*CoMoDaVideo*) class, then registering this *CoMoDaVideo* class to the specialized *ModeORM* class. As the data structure has changed, a database migration was required to synchronize the data structure with the database structure.

Table 6 LDOS CoMoDa dataset attributes mapped to MoRe data model

Category	Aspects	Supported in
User	Id	User.Id
	Age	User.Demographic.Birthday
	Gender	User.Demographic.Gender
	City	User.Contact.Address.City
	Country	User.Contact.Address.Country
	EndEmotion	User.Emotions.EmotionalState (*list*)
	DominantEmo	User.Emotions.EmotionalState (*list*)
	Mood	User.Mental.Mood
	Physical	User.Physiology.PhysiologicalState (*list*)
Context	Id	Context.Id
	Time	Context.Time.TypeOfDay
	DayType	Context.Time.DayType
	Season	Context.Time.Season
	Location	Context.Place.PlaceType
	Weather	Context.PhysicalCondition.Weather
	Social	Context.SocialRelation.SocialRelation (*list*)
	Decision	*not directly supported
	Interaction	*not directly supported
Item	Id	Video.Id
	Title	Video.Title
	Director	Video.Director
	Country	Video.Conuntry
	Language	Video.Language (*list*)
	Year	Video.Year
	Genre (3)	Video.Genre (list)
	Actor (3)	Video.Actor (*list*)
	Budget	Video.ItemAttributes (*list*)
	ImdbUrl	Video.ItemAttributes (*list*)

*denotes attributes not directly supported in the model, some adaptation was required

Discussion

The selected dataset can be seen as having a small number of features (32), but most of the existing dataset for CARS have fewer features (e.g. InCarMusic [41] as 18, DePaul Movie [28] has 6, and Trip Advisor [39] has 9). Even though, being able to use a bigger dataset is an interesting challenge that would allow us to test more attributes of the model, having used LDOS CoMoDa dataset allow us to test the ability of the model to store data, to maintain the relationship between the data categories (user, context, and item), and the ability to serve back the stored data.

The age attribute contained in the dataset was considered supported even when it was not stored directly as an integer, rather the model stores the birthday that which can be easily converted into age with a little processing. Maybe such processing of converting birthday into age should be directly included in the model in a future release of MoRe. The emotions (endEmotion and dominantEmo) and mood are considered as part of the user information, but as they are intended to reflect the user's state of mind in a specific situation, they are linked to the context, as the

MoRe models support to assign a specific context to emotional situation, meaning that the stored emotion was triggered/reading from the specified (if any) context.

4.2.2 Contextual Recommendation Feature

To test the ability of MoRe framework to generate context-based recommendations, which is the core functionality of the framework, the following use case was implemented.

Process

The process for this experiment consisted of using the framework's dataset generator to create the data in the database into a data matrix that can be fed to the algorithms, to generate recommendations, simulating the process that will be followed in a real-world CARS implementation.

The data used for this experiment was LDOS CoMoDa dataset that was loaded onto the framework model (as described in previous section). Having the dataset created, it was used to generate predictions using the both, contextual filtering and contextual modeling techniques.

Typically, algorithms are evaluated on their *rating prediction* ability, using metrics like Mean Absolute Error (MAE) and Root Mean Square Error (RMSE), such metrics were used to evaluate MoRe algorithms, except for SLIM and BPR algorithms, are they only support Top-N *item recommendation*, and not rating prediction, these algorithms were evaluated using ranking metrics Area Under the Curve (AUC) and Recall for 5 elements (Rec5).

In this experiment, all the techniques and algorithms of MoRe were tested, including the 24 combinations of the three splitting approaches (UserSplitting, ItemSplitting and UISplitting) with the 8 traditional recommendation algorithms (SLIM, BPR, GlobalAverage, UserAverage, ItemAverage, SVD++, UserKNN and ItemKNN).

Results

The obtained results loading the LDOS CoMoDa dataset into MoRe, generating a dataset with such information and applying the recommendation algorithms are shown next. Figure 3 shows the RMSE and MAE values for the contextual modeling algorithms; then Table 7 shows the result of the combination of filtering techniques with the baseline recommender algorithms.

Discussion

This experiment was used as a proof of concept and allows us to test the functionality of one of the main features MoRe provides: generating recommendations over stored data. All the algorithms contained in MoRe were testes with the same dataset, yielding somewhat similar results. The better performing algorithm was *ItemAverage* when used with *ItemSplitting* approach. In general, the contextual

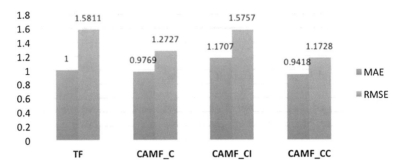

Fig. 3 Evaluation of contextual modeling algorithms on LDOS CoMoDa data

Table 7 Evaluation of contextual filtering algorithms on LDOS CoMoDa data

Technique	Algorithm	UserSplitting		ItemSplitting		UISplitting	
		AUC	Rec5	AUC	Rec5	AUC	Rec5
Ranking	SLIM	0.5522	0.0710	0.5507	0.6800	0.5526	0.0730
	BPR	0.5029	0.0030	0.5019	0.0030	0.5036	0.0039
		MAE	RMSE	MAE	RMSE	MAE	RMSE
Collaborative filtering	GlobalAvg	0.7700	0.9789	0.8509	1.0590	0.8509	1.0580
	UserAvg	0.7700	0.9789	0.7615	0.9757	0.7697	0.9788
	ItemAvg	**0.4728**	**0.7461**	**0.4870**	**0.0770**	**0.4969**	**0.7701**
Average	SVD ++	0.8573	1.0662	0.8561	1.0686	0.8552	1.0643
	UserKNN	0.8125	1.0447	0.8173	1.0501	0.8210	1.0502
	ItemKNN	0.7041	0.9272	0.7089	0.9331	0.7170	0.9388

Bold values are the best performing results

splitting approach yield slightly better results, especially when paired with *ItemAverage* baseline recommender.

The obtained results not necessary means that this will be the better performing algorithms in all cases, as these results are specific to recommend movies, and taking in consideration the specific characteristics of the LDOS CoMoDa dataset (user and context aspects considered, the data sparsity, and the number of instances). When recommending other elements, and using other contextual information, the algorithms will perform differently, fortunately, MoRe contains a large set of algorithms so developers can use the better performing one in their specific CARS.

5   Conclusions and Future Work

In this document, we have introduced MoRe, a comprehensive software framework to build context-aware recommender systems. To our knowledge, MoRe is the first CARS framework that provides developers a context-aware user model structure and implementation capable of organizing, storing and retrieving a fairly large set of user, context, and items aspects.

MoRe also provides a large set of state of the art recommending algorithms for both, traditional and contextual recommendation, covering the main recommending techniques existing in the literature. The main goal of MoRe is to serve as a tools for CARS developers, that can save the process of organize and create a complex structure that can support the information need by the system, and to save developers the process of implementing recommendation algorithms that can be complex and time consuming tasks, especially when developers what to test different algorithms over their data to choose the better performing one. MoRe is implemented in C# language, providing a large community of developers an easy to use tool to implement complex CARS systems. Our proposal follows Microsoft and object-oriented design principles and is published as an open source project so anyone can adapt it to their specific needs, or contribute to the improvement of the framework quality.

The framework proposal was compared against other proposals that contain similar feature regarding context modeling, or contextual recommendations, and showed to be the single proposal that packs most features. While other proposal focuses on proving very specific functionality like group-based recommendation or traditional recommendation algorithms, MoRe covert their functionality and much more, which helps MoRe stands out of other options, providing developers a larger set of features that can improve their productivity while implementing a CARS.

To test the functionality of MoRe's features, two use cases were implemented, first a real-world dataset was loaded into MoRe's data model, then, the data was exported from the model and fed into the recommendation algorithms, which showed that with a few lines of codes, this complex tasks were set up and working properly.

While the evaluations allowed us to demonstrate that MoRe work as expected, they are limited, and we believe that more elaborated experiments are needed to truly demonstrate the advantages that using MoRe in a real CARS development will bring to the table. Therefore, the next step in the process should be to present MoRe with expert and novice developers to gather their insights, especially with developers with no experience implementing CARS, as they will get the most benefit from the functionalities MoRe offers. Certainly, this would open a multitude of research questions and paths.

Another feature goal is to test the performance and reliability of recommendation algorithms in a more complex implementation while receiving a lot of requests simultaneously, as a real-world recommending system should support.

Acknowledgements This research is supported by the Maestría y Doctorado en Ciencias e Ingeniería (MYDCI) program offered by Facultad de Ciencias Químicas e Ingeniería attached to Universidad Autónoma de Baja California and for the Consejo Nacional de Ciencia y Tecnología (CONACYT) CVU 341714.

References

1. Chen B, Yu P, Cao C, et al (2015) ConRec: a software framework for context-aware recommendation based on dynamic and personalized context. In: Computer Software and Applications Conference (COMPSAC), 2015 IEEE 39th Annual. pp 816–821
2. Shi Y, Lin H, Li Y (2017) Context-aware recommender systems based on item-grain context clustering. In: Peng W, Alahakoon D, Li X (eds) Proceedings of AI 2017: advances in artificial intelligence. 30th Australasian Joint Conference, Melbourne, VIC, Australia 19–20 August, 2017. Springer International Publishing, Cham, pp 3–13
3. Campos P, Fernández-Tobías I, Cantador I, Díez F (2013) Context-aware Movie Recommendations: An Empirical Comparison of Pre-filtering, Post-filtering and Contextual Modeling Approaches. In: Huemer C, Lops P (eds) In International Conference on Electronic Commerce and Web Technologies. Springer Berlin Heidelberg, Berlin, Heidelberg, pp 137–149
4. Hawalah A, Fasli M (2014) Utilizing contextual ontological user profiles for personalized recommendations. Expert Syst Appl 41:4777–4797. https://doi.org/10.1016/j.eswa.2014.01.039
5. Hussein T, Linder T, Gaulke W, Ziegler J (2014) Hybreed: a software framework for developing context-aware hybrid recommender systems. User Model User-adapt Interact 24:121–174. https://doi.org/10.1007/s11257-012-9134-z
6. He C, Parra D, Verbert K (2016) Interactive recommender systems: a survey of the state of the art and future research challenges and opportunities. Expert Syst Appl 56:9–27. https://doi.org/10.1016/j.eswa.2016.02.013
7. Berkovsky S, Kuflik T, Ricci F (2008) Mediation of user models for enhanced personalization in recommender systems. User Model User-Adapted Interact 18:245–286. https://doi.org/10.1007/s11257-007-9042-9
8. Mettouris C, Papadopoulos GA (2016) Using appropriate context models for CARS context modelling. In: Kunifuji S, Papadopoulos AG, Skulimowski MJA, Janusz K (eds) Knowledge, information and creativity support systems: selected papers from KICSS' 2014. 9th International Conference, held in Limassol, Cyprus, on 6–8 November 2014. Springer International Publishing, Cham, pp 65–79
9. Adomavicius G, Jannach D (2014) Preface to the special issue on context-aware recommender systems. User Model User-adapt Interact 24:1–5
10. Gasparic M, Murphy GC, Ricci F (2017) A context model for IDE-based recommendation systems. J Syst Softw 128:200–219. https://doi.org/10.1016/j.jss.2016.09.012
11. Inzunza S, Juárez-Ramírez R, Jiménez S (2017) User modeling framework for context-aware recommender systems
12. Schilit B, Adams N, Want R (1994) Context-aware computing applications. In: First Workshop on Mobile Computing Systems and Applications, 1994. WMCSA 1994, pp 85–90
13. Dey AK, Abowd GD (1999) Towards a better understanding of context and context-awareness. Comput Syst 40:304–307. https://doi.org/10.1007/3-540-48157-5_29
14. Siolas G, Caridakis G, Mylonas P, et al (2013) Context-aware user modeling and semantic interoperability in smart home environments. 8th Semantic and Social Media Adaptation and Personalization, pp 27–32. https://doi.org/10.1109/SMAP.2013.19
15. Yurur O, Liu CH, Sheng Z et al (2016) Context-awareness for mobile sensing: a survey and future directions. IEEE Commun Surv Tutorials 18:68–93. https://doi.org/10.1109/COMST.2014.2381246
16. Brézillon P (2002) Modeling and using context: past, present and future. Rapp Rech du LIP6. Univ Paris 6:1–58
17. Perera C, Zaslavsky A, Christen P, Georgakopoulos D (2014) Context aware computing for the internet of things: a survey. IEEE Commun Surv Tutorials 16:414–454. https://doi.org/10.1109/SURV.2013.042313.00197

18. Zhang D, Huang H, Lai CF et al (2013) Survey on context-awareness in ubiquitous media. Multimed Tools Appl 67:179–211. https://doi.org/10.1007/s11042-011-0940-9
19. Baldauf M, Dustdar S, Rosenberg F (2007) A survey on context-aware systems. Int J Ad Hoc Ubiquitous Comput 2(4):263–277
20. Sabagh AAA, Al-Yasiri A (2013) GECAF: a framework for developing context-aware pervasive systems. Comput Sci Res Dev. https://doi.org/10.1007/s00450-013-0248-2
21. Brickley D, Guha RV (2000) Resource description framework (rdf) schema specification 1.0. In: W3C
22. Booch G, Rumbaugh J, Jacobson I (1996) The unified modeling language for object-oriented development. Unix Rev 14:29
23. Ormfoundation.org (2017) The ORM Foundation. www.ormfoundation.org. Accessed 8 Apr 2017
24. Gantner Z, Rendle S (2011) MyMediaLite: a free recommender system library. In: Proceedings of fifth ACM Conference on Recommender Systems, pp 305–308. https://doi.org/10.1145/2043932.2043989
25. Gantner Z, Rendle S, Schmidt-Thieme L (2010) Factorization models for context-/time-aware movie recommendations. In: Proceedings of the workshop on context-aware movie recommendation. ACM, New York, NY, USA, pp 14–19
26. Liu Y, Wang B, Wu B, et al (2016) CoGrec: a community-oriented group recommendation framework. In: Che W, Han Q, Wang H, et al (eds) Social computing: second international conference of young computer scientists, engineers and educators, ICYCSEE 2016, Harbin, China, 20–22 August 2016, Proceedings, Part I. Springer Singapore, Singapore, pp 258–271
27. del Carmen Rodríguez-Hernández M, Ilarri S (2014) Towards a Context-Aware Mobile Recommendation Architecture. In: Awan I, Younas M, Franch X, Quer C (eds) Proceedings of Mobile web information systems. 11th International Conference, MobiWIS 2014, Barcelona, Spain, 27–29 August 2014. Springer International Publishing, Cham, pp 56–70
28. Zheng Y, Mobasher B, Burke R (2015) CARSKit: a java-based context-aware recommendation engine. In: Proceedings of the 15th IEEE International Conference on Data Mining Workshops. IEEE, NJ USA
29. Abbar S, Bouzeghoub M, Lopez S (2009) Context-aware recommender systems: a service oriented approach. In: Proceedings of the 3rd International Workshop on Personalized Access, Profile Management and Context Awareness in Databases
30. Aguilar J, Jerez M, Rodríguez T (2017) CAMeOnto: context awareness meta ontology modeling. Appl Comput Informatics. https://doi.org/10.1016/j.aci.2017.08.001
31. Adomavicius G, Tuzhilin A (2008) Context-aware recommender systems. Proceedings of the 2008 ACM conference on Recommender systems, p 335. https://doi.org/10.1145/1454008.1454068
32. Meier J, Hill D, Homer A, et al (2009) Microsoft Application Architecture Guide
33. Troelsen A, Japikse P, Troelsen A, Japikse P (2015) ADO. NET Part III: Entity Framework. C# 60 the NET 46 Framew 929–999
34. Zheng Y (2015) A User's Guide to CARSKit. pp 1–7
35. Guibing Guo, Jie Zhang ZS and NY-S (2015) LibRec: A Java Library for Recommender Systems. Proc 23rd Conf User Model Adapt Pers 2:2–5
36. Ning X, Karypis G (2011) SLIM : Sparse Linear Methods for Top-N Recommender Systems. pp 1–10
37. Ricci F (2011) First International Workshop on Decision Making and Recommendation Acceptance Issues in Recommender Systems (DEMRA 2011) and Second International Workshop on User Models for Motivational Systems : the affective and the rational routes to persuasion (UM
38. Bobadilla J, Ortega F, Hernando A, Gutiérrez A (2013) Recommender systems survey. Knowl based Syst 46:109–132. https://doi.org/10.1016/j.knosys.2013.03.012
39. Zheng Y, Mobasher B, Burke R (2014) Context recommendation using multi-label classification. In: Proceedings of the 13th IEEE/WIC/ACM International Conference on Web Intelligence (WI 2014)

40. Karatzoglou A, Amatriain X, Baltrunas L, Oliver N (2010) Multiverse recommendation: n-dimensional tensor factorization for context-aware collaborative filtering. In: Proceedings of the 4th ACM Conference on Recommender Systems, p 79. https://doi.org/10.1145/1864708. 1864727
41. Baltrunas L, Kaminskas M, Ludwig B, et al (2011) Incarmusic: context-aware music recommendations in a car. In: E-Commerce and Web Technologies. Springer, pp 89–100
42. Baltrunas L, Ludwig B, Ricci F (2011) Matrix factorization techniques for context aware. Proceedings of the fifth ACM conference on Recommender systems, pp 301–304. https://doi.org/10.1145/2043932.2043988
43. Košir A, Odic A, Kunaver M et al (2011) Database for contextual personalization. Elektroteh Vestn 78:270–274

Executive Functions and Their Relationship with Interaction Design

Andrés Mejía Figueroa, J. Reyes Juárez Ramírez
and David Saldaña Sage

Abstract Human Factors have been for several decades one of the main factors of contention when trying to develop a usable system. From physical to cognitive characteristics and everything in between, the attributes of a user can impact on several aspects of the design of said software. Although there are guidelines for some characteristics there is no definitive model with what characteristics to consider, what metrics to use and their effect on the interface. In this chapter we talk about the effect of some of the executive functions (working memory and cognitive flexibility) depending on the interface design pattern used, and the relationship with the cognitive load produced by the design pattern.

1 Introduction

Computers today have changed the way we live, in some way or another ingrained in our everyday activities, from work to leisure, having access to news and information in an instant in the palm of our hand. But it wasn't always that way. In the early days of computing, computers were huge, extremely expensive, and needed specialized training in order to used them.

Along the way computer were being used in more and more situations and applications, from military to commercial, but with a lot of challenges. One being the Software Crisis, coined in 1968, which refers to the rising complexity of producing software, giving birth to the discipline of Software Engineering [1].

A. Mejía Figueroa (✉) · J. R. Juárez Ramírez
Facultad de Ciencias Químicas e Ingeniería, Universidad Autónoma de Baja California,
Tijuana, Baja California, Mexico
e-mail: mejia.andres@uabc.edu.mx

J. R. Juárez Ramírez
e-mail: reyesjua@uabc.edu.mx

D. Saldaña Sage
Facultad de Psicología, Universidad de Sevilla, Seville, Andalucia, Spain
e-mail: dsaldana@us.es

© Springer International Publishing AG, part of Springer Nature 2018
M. A. Sanchez et al. (eds.), *Computer Science and Engineering—Theory and Applications*, Studies in Systems, Decision and Control 143,
https://doi.org/10.1007/978-3-319-74060-7_2

25

Another problem which slowly emerged but was being noted was the problem of usability of software and physical systems. Bad design choices caused several fatal errors and accidents because of user error that could have been prevented with a good design [2]. As such the discipline of Human-Computer Interaction (HCI) emerged, formed by several other disciplines, such as human factors, cognitive science, and psychology.

HCI first appeared as the application of cognitive science in the information technology scene. Incidentally, as personal computing took off in the late 70s, so did cognitive science as a way to model the mind and its processes [2]. Thus, laying the groundwork of HCI by presenting the necessary tools needed to tackle the slew of new challenges presented with universal access to personal computing.

One of the biggest contributors to HCI was the discipline of Human Factors and Ergonomics (HFE), which focuses on human characteristics and its implication on the design of any device and system. Many of the HFE contributions in industry and ergonomics made its way into HCI, such is the case with cognitive ergonomics (or cognitive engineering), which studies cognition in a work environment; this includes things such as perception, mental workload, decision-making, stress, among others.

Some examples of cognitive engineering include the Human-Processor Model and the GOMS family of models proposed by Stuart K. Card, Thomas P. Moran and Allan Newell in 1983. Described in their seminal work: The Psychology of Human-Computer Interaction [3], in which they propose and describe, a cognitive model used to assess the usability of a software system and the time to complete a task in that software. They divided the human factors into three subsystems: perceptual, cognitive and motor, all which are needed to interact with any system properly. In each subsystem they outlined time frames for each process in order to calculate and estimated time of completion of a specific task. The final completion time consisted of the sum of the time of all the processes involved of each subsystem in the task. This model served as a base for future cognitive architectures and subsequently modified versions of the GOMS model.

So, in order for HCI to achieve its goal of usable interfaces, it must consider not only computer factors (software and hardware) but also the task and all the human factors that could have an effect in achieving the completion of the task at hand [4].

As a result of research in HCI, there exist several guidelines for the design of interfaces for users with some form of special needs of varying characteristics, such as the Web Accessibility Initiative by the W3C [5] to make web content more accessible to people with some form of disability. Another example is research in the area of online tutoring and the use of information and communication technologies in education [6], on how to properly teach a person new knowledge and skills using those tools, in many cases without a teacher or tutor present.

The most common way of interacting with software systems is through a Graphical User Interface (GUI), as previously mentioned, there are cognitive models which help predict user behavior when using a GUI, but there is still much to do, since there are other factors that affect usability. One of which is the cognitive load produced by the design of the interface.

One of the mayor aspects of consideration when designing a usable GUI is one of the golden rules of HCI: Reduce cognitive load [2]. The way we normally try to achieve this is by simplifying the number of items on the interface to the bare minimum so we don't overload the user with information and choice. Much of the nature of cognitive load is determined by many factors, one of which are our cognitive capabilities, including some of our executive functions.

There is currently no metric for the cognitive load produced by an interface design, although it can be inferred through other means. Every aspect of the interface design consumes mental resources, thus causing a certain degree of cognitive load, thus the choice of interface design patterns, depending on the task to be done, can affect the level of cognitive load induced and in turn affect the usability of the software.

In this chapter, we present advances pertaining our cognitive capabilities, mainly executive functions and their relationship with interface design, we present some interaction rules based on measurements of executive functions and some of the most common graphical interface design patterns, obtaining specific rules for GUI adaptations for varying levels of some cognitive capabilities, in this case, the executive functions of working memory and cognitive flexibility, and analyzed the influence of the cognitive load produced by the interface design patterns.

2 User Modelling and Interface Design

Modelling the user has become one of the more important parts of the software development life-cycle. With the current trend of User-Centered Design and User Experience Design, developing software is not only reserved to software engineers, now it is needed team members with training in other disciplines and/or other expertise to properly analyze and develop a more usable and friendly system.

As mentioned earlier HCI developed from a combination of disciplines, from Human Factors, to cognitive science. One of the first contributions of HCI was an attempt to model the way a person thinks and behaves in order predict how a person would use a system. As a result, there are several cognitive architectures in HCI as a means to model user behavior and interface evaluation. Byrne [7] defines a cognitive architecture as a hypothesis about aspects of human cognition that remain constant over time and are independent of the task. As such we can such those approximations of the real human cognition to have a better understanding on the proper design and analysis of software interfaces.

One of the milestones of cognitive modelling in HCI was the Model Human Processor, proposed by Card et al. [3], where it was the first that introduces the notion on information processing from the user and qualitative and quantitative predictions of user actions based on psychological studies. It also gave rise to the GOMS family of cognitive architectures which are some of the most widely used today.

GOMS (Goals, Operators, Methods, and Selection Rules) is a derivation of the Model Human Processor proposed by Card et al. [3]. The GOMS model functions in a similar way to a task analysis by analyzing the actions needed to perform a task in an efficient manner. The GOMS model has several components, them being:

- Goals. What the user wants to accomplish.
- Operators. Actions needed to perform each goal, such as keyboard strokes, button presses, etc.
- Methods. Sequences of operators needed to accomplish said goal
- Selection Rules. Conditions where the user might select one method over another.

With GOMS is also possible, much like the Human Model Processor, to calculate an average time of completions of a task based on average timing of basic actions, such as a button press or move cursor, obtained from previously done research.

Subsequent variations of the GOMS model were presented, such as the KLM (Keystroke-level Model), presented by Card et al. [3], where the task analysis reaches keystroke level operators, such as key presses. Execution time is the emphasis ignoring all else, such as goals, methods and selection rules. It is considered a simplified version of GOMS. And NGOMSL proposed by Kieras [8], where a formalized version of GOMS is presented with structured rules and notations.

Another popular architecture, EPIC (Executive-Process Interactive Control), proposed by Kieras et al. [9], where executive processes are represented that control other processes during a multitasking performance. Also, perception and motor behaviors are given their own module. Procedures and production rules specify modules, thus requiring knowledge about the task to be done.

The ACT-R (Adaptive Control of Thought-Rational) architecture, proposed by Anderson [10], was first developed as a theory of higher cognition and learning but subsequently adapted for HCI. It assumes that knowledge can be classified in two kinds: declarative and procedural. Declarative knowledge is represented in the form of chunks and are saved and made accessible through buffers. The most recent version, ACT-R/PM, proposed by Byrne [11], incorporates perceptual motor capabilities similar to the EPIC architecture.

Another type of area where human factors take center stage in development of the software is in Adaptive Interfaces. Adaptive Interfaces, as the name implies, adapt to the user's characteristics and the context of use of software, providing, in theory, a more personalized and usable experience [12].

One of the core parts of the architecture of an Adaptive Interface is the User Model. In order to develop an adaptive system, we must first determine what will be the user characteristics that will considered thus composing the user model. With this model, the user interface can be adapted based on the values, rules, and guidelines that resulted of user research. Most software applications that use user models often just consider some aspects of the user that they deem relevant to the application, there is no generic solution to be used, although there is research towards achieving that goal.

One example is the AVANTI project [13] in which an AUI was developed for a web documents, adapting itself to the users to some extent based on some information submitted by the user, such as language, eyesight, motor abilities, language and experience with similar applications. It also adapts itself in a dynamic way as they interact with the system based on user familiarity with specific tasks, ability to navigate, error rate, disorientation, user idle time, and repetition of interaction patterns. The adaptation also supports Multi-Modal systems by featuring integrated support to various input and output devices with appropriate interaction techniques in order to accommodate users with disabilities. The UI supports users with light to severe motor disabilities and blindness.

In Gajos et al. [14] is presented an automatic personalized user interface generator called SUPPLE. With SUPPLE is possible to generate interfaces adapted to the user's devices, tasks and preferences. Also, there is case study presented where in can also consider some user characteristics, in this case motor skills, by developing a usage model based on users with motor deficits. They define interface generation as a optimization problem and demonstrate that is computationally feasible. Adaptation is based on a device model for the limitations of the device for which an interface will be generated and a usage model to represent user preferences, in conjunction with a cost function that is used to calculate the current effort of use.

Although cognitive characteristics are being considered there are still several that there is little to no research on the impact they could have on interface design, some of those being executive functions. One of the most commonly considered is working memory, since it serves as a component of several other cognitive functions, but others such as planning and cognitive flexibility have had drawn very little attention.

3 Executive Functions

Executive function is a term used to describe cognitive processes necessary for the completion of goal-oriented tasks, usually carried out in large part at least, in the frontal lobes.

As many terms and definitions in science and engineering, executive function defies a formal definition for the reason that as more research is done on the nature of executive functioning, sometimes contradictory results are presented generating controversy and discourse. Even with similar yet sometimes different definitions, practically all agree that executive functions are the set of cognitive processes where cognitive abilities are used in goal-oriented tasks [15].

Delis [16] defines executive functions as the ability to manage and regulate one's behavior to achieve the desired goal. Similar to Delis, Miller and Cohen [17] suggest that Executive Control involves the cognitive abilities needed to perform goal-oriented tasks. Lezak [18] describes executive functioning as a collection of interrelated cognitive and behavioral skills that are responsible for the goal-directed activity, includes intellect, thought, self-control, and social interaction.

Notions of executive functioning were first observed in studies related to patients with abnormalities or injuries in the prefrontal cortex such as in Pribam [19] and Luria [20], which noted impaired ability to evaluate and regulate their behaviors and goal oriented mindset.

Some of the elements of executive functioning include short and sustained attention, task initiation, emotional regulation, working memory, cognitive flexibility or shifting, planning and problem-solving ability. This encompasses not only cognitive processes but also emotional responses and behaviors [21]. For this reason, Executive functions are classified in "hot" or "cool" processes, where "hot" processes are the ones that have an affective component, such as emotional regulation, and "cool" processes which involve purely cognitive functions, such as working memory [21].

Further research pertaining the prefrontal cortex and executive functions was done achieving similar observations, but subsequent studies also reproduced contradicting results. Such is the case in Godefroy et al. [22] where the notion that all the processes for executive functions were in the frontal lobes was put in doubt. A study showed that patients with lesions of the prefrontal or posterior cortices were submitted to a series of conflicting and combined tasks. The results showed the expected prominent role of the frontal lobe but also that executive functions depend on multiple, separate, and modular control processes because certain patients with frontal lobe injury performed well on tests designed to assess executive functioning while others did not.

These revelations make sense since the prefrontal cortex is shown to be dependent on outgoing and ingoing connections to all other brain regions, such as the occipital, temporal and parietal lobes. Damage or deficits on any related region would result in some executive function impairment. The prefrontal cortex works as a hub of communications between all regions of the brain that pertain to executive functioning, if communication between those regions is lost, or the region is damage, impairment in executive function occurs.

Symptoms of executive dysfunction include the inability to maintain attention, lack of impulse control, low working memory capacity, inability to plan future activities, inability to shift attention between different tasks or stimuli, difficulty generating new knowledge, among others.

As we can see, although there is some controversy on the nature of executive functioning and the great number of definitions that have been proposed, there is the consensus that it involves the cognitive processes that manage goal-directed behavior.

3.1 Executive Functions and Interaction Design

A User Interface (UI) is defined as the point where information exchange between the user and the system occurs [23]. Today, with most personal computers and mobile devices such as tablets and smartphones, GUI's are still the preferred type of interface for the average user.

GUI's normally consist of several interface components, such as buttons, labels, text areas, for example. Depending on the type of task that needs to be accomplished, there can be a wide selection of interface components on which the user can accomplish the task. The recurring use of a set of interface components for a particular task is called User Interface Design Patterns [23]. For example, one of the most common patterns is the login screen. Usually, it consists of two textboxes, one for your login and other for your password, and two buttons, one to enter the system and other to cancel. Other examples of design patterns include Tabs, Wizard, Accordion, and Navigational Drawer just to name a few.

Often during software development, depending on the results of the user and task analysis, certain usability patterns and interface design decisions are made in the hopes that the interface will be usable for most of its intended user audience, nevertheless often it is not the case.

Some patterns might be useful for certain tasks, but require certain user characteristics to be properly used, limiting the usability for some users that might have an impairment or notably different characteristics.

3.2 The Problem of Cognitive Load

One of the main problems that HCI tries to tackle in interface design is to minimize as much as possible the cognitive load upon the user caused by the design decisions of the user interface.

Cognitive load is defined as the mental resources available for the task at hand [24]. The capacity of mental resources varies from person to person and has several factors that contribute to it, but is more tied to working memory and long-term memory, being working memory one of the most important executive functions since is used as a component of other more complex executive functions, such as planning and cognitive flexibility [25].

Cognitive load can be classified into three categories depending on the nature of the load [24]. Some load is imposed on the working memory by the nature of the complexity of the task, called intrinsic load. Another category of cognitive load takes place from the way the information or task is presented to the user. In the case of interface design, that means that the design decisions of the interface have a direct impact on this type of cognitive load, which is called extraneous cognitive load. Both types of the cognitive load must be dealt with by the resources available to the working memory allocated to both cognitive loads, this type of load is called germaine load, which are the resources needed for learning and storing information in schemas in long-term memory.

Cognitive load can never be fully eliminated from any task since we always need mental resources for any type of action, be it a movement, a selection or simply searching for an item on screen; but there are some steps to minimize it.

In the case of interface design, we can attack the problem by minimizing the extraneous load caused by the difficulty of use of the interface and quantity of information shown at once, and in some cases the intrinsic load, by simplifying and automating some steps of the task.

3.3   Can We Measure Cognitive Load?

One of the leading researchers in cognitive load theory, John Sweller in his work [26] lists several ways in which we can measure cognitive load, although there is no exact metric for cognitive load, there is way we can infer the effects of cognitive load and have an approximation of a measurement.

3.3.1   Indirect Measures

The first approximations of a measurement for cognitive load were developed by indirectly by experimentation that examining the relationship between problem-solving and learning. One of the approximations was with the use of computer models. It started with research focused on the inefficiency of problem-solving ability as a learning strategy where it was demonstrated that learning strategies that used a lot of problems solving search led to worse outcomes than strategies with less problem-solving search. They argued in [26] that the problem-solving searches causes high extraneous cognitive load, impeding proper schema creation and acquisition.

With a production system model Ayres and Sweller [27] demonstrated that higher problem-solving search required a much more complex model to be simulated, thus giving credence that it would be a much higher burden to the working memory.

Another approximation to infer cognitive load was the use of performance indicators during the acquisition or learning phase. Chandler and Sweller [28] showed that students using a learning technique that increased cognitive load impacted performance during the learning and acquisition phases. In later research it also showed that it also increased error rates were higher during the acquisition phase with high cognitive load learning techniques were used, with this in mind, subsequent research gave credence to the idea that error rates might also be used as an indirect way of measuring cognitive load [29] since results showed that students made the most errors in mathematical tasks that required high decision-making skills with many variables to be considered.

3.3.2   Subjective Measures

Other ways to get an approximation of cognitive load is using subjective measures. Previous research [30] indicated that user introspection of mental effort could be used as a way of the index of cognitive load. Mental Effort is defined as an aspect of cognitive load that refers to the cognitive capacity that is allocated to accommodate demands imposed by the task, thus can be considered a reflect of cognitive load.

In [31] a 9 point Linkert Scale was used, which ranged from 1, which represented low mental effort, and nine which represented high mental effort. The results showed a correlation between students that used hypothesized lower cognitive load instructional design and the lower mental effort rating that they gave; the same with the students that used a higher hypothesized cognitive load instructional design, which yields higher mental effort scores.

3.3.3 Measurement Through a Secondary Task

The traditional method for the measurement of cognitive load is through the use of a secondary task [32] in what is called a dual-task methodology. It requires that the subject engages in a subsequent task after a primary one. Depending on the cognitive load that the subject suffers from the first task, his performance will suffer in the second one because of it.

3.3.4 Physiological Measures

One of the first physiological measures of the cognitive load was of cognitive papillary response [33]. By testing several tasks with the very varied load to working memory found that there was a relationship between pupil dilation and perceived memory load. Pupil dilation increased with rising cognitive load measures, although there are some age limitations, the dilations decreases with age, in the study elderly participants did not show pupil dilation on cognitive tasks.

Another successful way of measuring cognitive load in a more accurate and precise way than with subjective measurements is with the use of Functional Magnetic Resonance Imaging (fMRI) and Electroencephalography (EEG) [34, 35]. During an experiment where a subjective measure of mental effort was used in a task, an EEG captured alpha, beta and theta brain wakes. It showed that even when in the subjective scores there was no discernable difference between tasks, the EEG measurements were sensitive enough to show a sizeable difference in perceived mental effort.

Other forms of measuring cognitive load are the use of eye tracking in multimedia environments and applications [36]. It was found that different combinations of text and images required a different level and cognitive processes, thus more cognitive load; which were correlated with varying degrees of eye fixation.

Although there are some ways to get an approximation of the Cognitive Load that a user is currently having, there is no real way to measure the induced Cognitive Load of a User Interface on the user, or more precise design guidelines for different levels of Executive functions, such is the case of working memory, which is greatly affected by cognitive load.

Different combination of GUI components and Interface Design Patterns can cause varying levels of Cognitive Load depending on the user's mental capacities, which can vary greatly. Adapting to these very different and varying capacities can be a daunting task, making designing a usable software across a varied user base extremely difficult to achieve, such is the case with specialized software for users with special needs.

As we can see, there has been work on improving usability by integrating HCI and usability engineering practices in the software development life cycle. Frameworks were proposed for the generation of usable user interfaces, mostly focusing on the tasks that the user must accomplish with some integrating some user modelling aspects. There is significant advancement towards a standardized user model but there is still much work to be done before we can have a generic solution with a standard rule set for each user characteristic. Also, cognitive load continues to be one of the main problems when designing a usable interface, without an easy to use metric or a cognitive load score to interface design patterns it is hard to determine what level of cognitive load it produces to the user, since cognitive load depends on several factors including the user's own cognitive capabilities.

To better define what to do with different user characteristics we conducted a study where we obtained interaction rules based on measurements of executive functions in order to better choose interface design patterns based on the executive function measurements and usability testing metrics with a wide variety of users. While also analyzing the data to see the relationship between the cognitive load produced by each interface pattern and the executive functions.

4   Research Questions

We had two main research questions:

1. What Interface Design Patterns are better suited for different levels of Executive functions (In this case working memory and cognitive flexibility)?
2. What is the relationship between Cognitive Load caused by User Interface Design Patterns and the different levels of working memory and cognitive flexibility?

5   Experimental Design

5.1   Objectives

The purpose of the experiment is to determine if there is a relationship between executive functions measurements (working memory and cognitive flexibility), GUI Design Patterns and Cognitive Load with the use of neuropsychological tests, while also determine possible Interaction Rules based on the measurements of Executive functions and usability testing results.

5.2 Sample

Group of 105 children, ages 7–12, composed of three subgroups: 35 children with Autism Spectrum Disorder (ASD) level and 35 children with ASD level 2 according to the DSM-5 [37]; and 35 typical children. Children with ASD were chosen because of the impairments in the key Executive functions being examined as a symptom of Autism [38].

5.3 Structure of the Study

The experiment is divided in three main phases:

- Phase 1: Application of all neuropsychological tests to the children needed to evaluate the executive functions of working memory and cognitive flexibility.
- Phase 2: Usability testing of eight prototypes of an Augmentative and Alternative Communication (AAC) app using the Picture Exchange Communication System (PECS), each one using a different interface design pattern. The interface design patterns selected were: Accordion, carrousel, scrolling menu, navigational drawer, tabs, one-window drilldown, two-panel select, and wizard. The task was the same for each one of the prototypes, form a three-part sentence, for example, I want to play.
- Phase 3: Analysis of the data, with different objectives. The first objective if the obtention of a ruleset using Classification Trees with the target being the user interface design pattern based in the executive function measurements, usability testing results are used here. The second objective is find out if there is a direct influence of the executive functions on the cognitive load produced when using each of the applications, to see if there is a meaningful statistical significance using regression analysis.

5.4 Variables

Measurements of the following were taken of all children:

- Working Memory. A temporary system where we can store and manipulate information in the short-term memory. The WISC-IV intelligence test was used for measurement.

- Cognitive Flexibility. Ability to shift to a different thought or action in response to a situation change. The neuropsychological battery NEPSY II was used to measure cognitive flexibility.
- Cognitive Load. Mental resources used by the task at hand. Measured by using a modified version of the NASA-TLX for children.

Usability testing metrics being considered:

- Binary success.
- Error rate.
- Time on task.
- Level of success.

Interface design patterns:

- Pattern 1: Accordion
- Pattern 2: Carrousel
- Pattern 3: Scrolling menu
- Pattern 4: Navigational Drawer
- Pattern 5: Tabs
- Pattern 6: One-Window Drilldown
- Pattern 7: Two panel select
- Pattern 8: Wizard

5.5   Instruments

5.5.1   Nepsy II

NEPSY II [39] is a set of neuropsychological test that is used in different combinations to evaluate the neurological development of children from the ages of 3–16 years in six domains. In this case, we are only going to center on the Executive function domain.

The Attention and Executive function domain is composed of the tests of Animal Sorting, Auditory Attention, Response Set, Clocks, Design Fluency, Inhibition, and Statue. These subcomponents evaluate several Executive functions, such as sustained and selective attention, set shifting or cognitive flexibility, planning, inhibition and auto regulation and monitoring.

The test that will be used from the NEPSY II Attention and Executive function domain are the ones related to cognitive flexibility or set shifting:

- Animal Sorting (AS). This test is designed to evaluate the ability to formulate basic concepts, transfer then into actions and change the focus of his attention. The participant classifies cards in two groups, each one with different criteria defined by the child.

- Auditory Attention (AA) and Response Set (RS). Auditory Attention evaluates selective auditory attention and to sustain it. Response Set assesses the ability to set shift and maintain new conditions by inhibiting previously learned responses and correctly responding to new stimuli. The child listens to a list of colors and touches the appropriate circle of color when he or she hears the target word.

5.5.2 Wisc-IV

WISC-IV [40] is an intelligence test administered to children between the ages of 6 and 16 years of age. It generates a general intelligence coefficient (IQ) and five primary IQs, of which for this study we will be focusing on the tests needed for the working memory index.

The working memory index is the one in charge of evaluating the storage and retention of information and its manipulation. It is composed of two subtests:

- Digit Span (DS). Analyzes short term memory, which gives us an idea of his sequencing abilities, planning ability and cognitive flexibility.
- Letter-Number Sequencing (LN). Analyzes memory retention capacity and the ability to combine different types of information, organize said information and elaboration of organized groups based on the same information.

5.5.3 Modified NASA-TLX Test for Children

Adapted version of the NASA Task Load Index for children [41] for measuring mental workload, also referred as cognitive load. NASA-TLX is a subjective assessment of mental workload of a perceived task divided into six subscales: mental demand, physical demand, temporal demand, performance, effort, and frustrations.

Usually the original NASA-TLX used Linkert scales for each of the subscales, ranging from very low to very high, with a range of values from 0 to 100 with increments of five. In the modified version, a physical ruler was used with a cartoon on each side of the scale representing both extremes of the subscale being measured, for example, for the mental demand subscale on the least demanding side a cartoon of a relaxed child can be seen doing homework, while on the other extreme a very tired cartoon of a student doing homework can be seen. It is also accompanied by an extended dialog and questions that the examiner must explain before each of the scales is responded by the child. Both test return a single value in the 0–100 range for mental workload of the given task.

5.6   Methodology

Measurements of working memory and cognitive load are taken using the WISC and NEPSY tests. The values obtained are not transformed and standardized for their age according to the normal procedures of evaluation using those tests. The values used are the number of right answers of each tests in order to properly observe the variance of the results between the subjects. If the results were standardized it would have been a result based on the age of each of the participants. For example: it is not the same a score of 12 in a test for a participant of an age of 7 years than that same score for a participant of 12 years of age.

After measurements of the executive functions are taken, the participants proceed to usability testing of eight prototypes of an AAC application using the PECS system on an Android tablet. The task is to form a simple three-part sentence in each prototype. Each session is videotaped for analysis and the usability metrics are obtained. The order of the testing was counterbalanced and spreaded out in different days depending on the availability of the participants. Each usability test took about 5 min per prototype.

After each usability test, the modified NASA-TLX test was applied and a cognitive load score was obtained for each prototype tested for each participant.

The neuropsychological tests and modified NASA-TLX test were applied by psychology students and trained personnel from the schools, and the usability testing carried out by interaction design students.

For the research question of which interface design patterns are better suited for different levels of working memory and cognitive flexibility, we used classification trees in order to obtain rules with the target on the best interface design pattern based on the results of the executive function testing.

The main dataset was divided into eight minor datasets based on the results of the usability testing, for example, one dataset consists of the best performing registry of every participant, another one of the second best performing registry, and so on until we reach the eight-dataset consisting of the worst performing registry of each participant.

Classification trees were used on each dataset and a ruleset was obtained, which was subsequently tested with a newer and smaller dataset consisting of 30 randomized children which did not participate in the composition of the original dataset.

To find the relationship of cognitive load and the executive functions we transformed the data and performed a repeated measures ANOVA and a regression analysis for each design pattern used.

5.7 Results

5.7.1 Data Analysis

The dataset obtained consisted of eight registries per participant, each registry made of the five results of the executive functions measurements, the cognitive load score, the interface design pattern, and the usability testing metrics, for a total of 840 registries.

All executive function measurements showed signs of high collinearity between them (Correlation of minimum 0.716 and a maximum of 0.971), and VIF values ranging from 2 to 5 depending on the regression being performed.

For the repeated measure ANOVA and the regression analysis it was necessary to perform data transformations in order to comply with the normality of residuals and the independence assumptions since the data failed the Shapiro-Wilks test and showed high correlation between variables. To deal with collinearity we used Principal Component Analysis on the executive function measurements which yielded one component for the five variables that explained 70.325% of the variance and for the normality of residuals we transformed the data by using Box-Cox transformation.

5.7.2 Interaction Rules

Classification trees were applied to the datasets, the following rules were simplified from all the rules obtained from the trees, only the trees where a decent accuracy level of prediction are analyzed since some of the obtained trees show too much variance and don't really give any useful information.

The tests for Auditory Attention (AA) and Response Set (SS) have a maximum score of 30. Animal sorting had no maximum value, but a value of around 8 is considered very high. Digits Span (DS) has a maximum score of 16, so do Numbers and Letters (NL).

From the best performing interface patterns:

IF AA < 25.5 THEN
(pattern1 4.3%, pattern2 6.4%, pattern3 6.4%, pattern5 8.5%, pattern7 72%, pattern8 2.1%)
IF AA >= 25.5 && DS >= 14.5
(pattern1 0%, pattern2 25%, pattern3 0%, pattern5 75%, pattern7 0%, pattern8 0%)
IF AA >= 25.5 && DS < 14.5 && SS<26.5
(pattern1 0%, pattern2 57%, pattern3 0%, pattern5 0%, pattern7 43%, pattern8 0%)
IF AA >= 25.5 && DS < 14.5 && SS>=26.5
(pattern1 0%, pattern2 11%, pattern3 5.6%, pattern5 33%, pattern7 5%, pattern8 0%)

In this case we can see an overwhelming better result for the pattern 7, this being Two-Panel Select. We can see that users with a score of 25.5 or lower of Auditory Attention (AA), which is one of the tests we used related to cognitive flexibility, performed a lot better with two-panel select, closely followed by the Tabs pattern when the users had a better AA score.

Other influential user attributes were digit span (DS) and response set (SS). With a decent digit span score over 14.5 the users improve usability scores on several other patterns such as tabs and carrousel. When considering set shifting users with a score of less of 26.5 performed better with tabs by a small margin (difference of 13%) instead of two-panel select. This ruleset showed a 53.33% on average accuracy when tested.

We can generalize with this tree that as a rule of thumb that users with a medium level to low auditory attention will perform better using the two-panel select, with tabs and carrousel closely behind when AA is a little higher than 25 and DS is around 14.

For the second best performing interface patterns:

IF DS < 1.5
        (pattern1 0%, pattern2 1.1%, pattern3 0%, pattern5 89%, pattern6 0%, pattern7 0%)
IF DS >=1.5 && DS <10.5 && NL < 5.5
        (pattern1 0.9%, pattern2 55%, pattern3 27%, pattern5 0%, pattern6 0%, pattern7 9.1%)
IF DS >=1.5 && DS <10.5 && NL >= 5.5
        (pattern1 0%, pattern2 18%, pattern3 36%, pattern5 27%, pattern6 9.1%, pattern7 9.1%)
IF DS >=1.5 && DS >=10.5 && SS < 29.5 && SS >=25.5
        (pattern1 0%, pattern2 17%, pattern3 0%, pattern5 67%, pattern6 0%, pattern7 17%)
IF DS >=1.5 && DS >=10.5 && SS < 29.5 && SS <25.5
        (pattern1 5%, pattern2 20%, pattern3 5%, pattern5 20%, pattern6 0%, pattern7 50%)
IF DS >=1.5 && DS >=10.5 && SS >= 29.5
        (pattern1 0%, pattern2 8.3%, pattern3 8.3%, pattern5 17%, pattern6 0%, pattern7 67%)

In the case of the second ones, we start to see more variance in the results. User with a low score of 1.5 in the digit span test performed better with the Tabs pattern in 89% of cases. When digit span is greater than 1.5 and numbers and letters less than 5.5 we see that Carrousel and Scrolling Menu have significant gains.

After this we see the influence of Response Set, when it has a somewhat high value between 25.5 and 29.5 again the most usable pattern being Tabs. The accuracy obtained on average was 41.37%.

From here we jump to the sixth, seventh and eighth place performing patterns trees, since the other ones showed very low accuracy in the low single digits.

For the sixth performing patterns we have the following ruleset:

IF AS >= 4.5
        (pattern1 21%, pattern2 3.7%, pattern3 16%, pattern4 19%, pattern5 1.2%, pattern6 21% pattern8 17%)
IF AS< 4.5 && NL >= 5.5
        (pattern1 32%, pattern2 4.9%, pattern3 7.3%, pattern4 17%, pattern5 2.4%, pattern6 17% pattern8 20%)

IF AS< 4.5 && NL < 5.5
>   *(pattern1 9.5%, pattern2 0%, pattern3 14%, pattern4 24%, pattern5 0%, pattern6 43% pattern8 9.5%)*

Here we see things more spread out since this is some of the least effective patterns. At this level users with a value above 4.5 for Animal Sorting (AS) used accordion, scrolling menu, navigational drawer, one-window drilldown and wizard quite evenly. When also considering NL > 5.5 we see a slight shift toward accordion and wizard. When the value of NL < 5.5 it shifted toward Navigational Drawer and one-window drilldown. The accuracy of the classification tree was on average 24.13%.

For the seventh level of interface patterns:

IF AS < 0.5
>   *(pattern1 71%, pattern2 0%, pattern3 29%, pattern4 0%, pattern6 0%, pattern7 0% pattern8 0%)*

IF AS >= 0.5 && AS < 2.5
>   *(pattern1 0%, pattern2 0%, pattern3 41%, pattern4 29%, pattern6 24%, pattern7 5.9% pattern8 0%)*

IF AS >= 0.5 && AS >= 2.5 && DS >= 15.5
>   *(pattern1 14%, pattern2 14%, pattern3 43%, pattern4 0%, pattern6 29%, pattern7 0% pattern8 0%)*

IF AS >= 0.5 && AS >= 2.5 && DS < 15.5 && AS >= 6.5
>   *(pattern1 22%, pattern2 0%, pattern3 0%, pattern4 56%, pattern6 0%, pattern7 0% pattern8 22%)*

IF AS >= 0.5 && AS >= 2.5 && DS < 15.5 && AS < 6.5
>   *(pattern1 22%, pattern2 0%, pattern3 0%, pattern4 56%, pattern6 0%, pattern7 0% pattern8 22%)*

Here we see that the difference between increments of the measurements and a noted change in best pattern for each user in this level. We see that users with very low Animal Sorting level, less than 0.5 at this level the Accordion pattern seemed to stand out. With a little more of Animal sorting and considering Digit Span, Scrolling Menu and Navigational Drawer also making some gains with a 43 and 56% respectably. The accuracy of the tree on average was 17.24%.

And finally, on to the worst performing patterns dataset.

IF AS < 4.5
>   *(pattern1 42.8%, pattern3 28.5%, pattern4 0%, pattern8 28.5%)*

IF AS >= 4.5
>   *(pattern1 0%, pattern3 0%, pattern4 1.5%, pattern8 98%)*

As can be seen in the case of the task that was considered the wizard pattern seemed to be the worst performing in most cases. The accuracy of this classification tree was a high 98.4%.

In summary, we can determine from the best performing classification tree, that the great majority of users with a medium to low AA (AA < 25.5) will perform better with two-panel select, followed closely by tabs when the user has a AA > 25.5, signaling that the tabs pattern might be slightly harder to use.

From the second best patterns we can see similar results but considering other metrics, being digit span the main one, since those with a near score of 0 in digit span performed better using tabs, which was one of the best evaluated in the tests. Once the value of digit span incremented above 1.5, we start to see some variance between carrousel, scrolling menu and tabs, with an influence in part of by response set.

The unexpected surprise was the performance of the wizard pattern, which was overall the worst one, at least for this task. One theory is that since the participants had deficiencies in working memory they had a hard time remembering the previous step and committed mistakes, unlike two-panel select, they had on all times the information on screen.

5.7.3 Analysis of Variance

Applying a repeated measures ANOVA, the data violates the assumption of sphericity, to correct this issue we use the Huynh-Feldt correction (Epsilon = 0.868).

With the correction the mean scores for cognitive load for each Interface Design Pattern were statistically significantly different with an $F (6.073, 364.36) = 64.355$, $p < 0.05$.

When considering the executive function component as a covariate we can see a significant interaction $F (6.073, 364.36) = 7.369$, $p < 0.05$.

5.7.4 Regression Analysis

To see the impact of executive functions on the cognitive load produced by each of the patterns, after doing all the necessary data transformations we proceeded to do a linear regression analysis, where the executive function component as the independent variable, and the cognitive load from the modified NASA-TLX.

For the accordion pattern the results of the regression analysis were (Table 1): The regression equation with the coefficients being:

$$\text{Cognitive Load} = 2.742 - 0.149\text{EFComponent}$$

Table 1 Regression results for the accordion pattern

R	R squared	Adjusted R square	Std. error of the estimate
0.369	0.136	0.122	0.3788

Table 2 Regression results for the Carrousel pattern

R	R squared	Adjusted R square	Std. error of the estimate
0.333	0.111	0.097	0.12354

For this analysis the cognitive load metric was transformed by a λ of 0.303 via Box-Cox transformation to achieve normality.

For the regression analysis of the carrousel pattern the results were (Table 2):
With the regression equation being:

$$\text{Cognitive load} = 1.477 - 0.43\text{EFComponent}$$

The cognitive load value was previously transformed with a λ of 0.1414 to achieve normality.

For the scrolling menu pattern, the regression analysis results were (Table 3):
The equation being:

$$\text{Cognitive load} = 26.799 - 9.027\text{EFComponent}$$

The analysis for the navigational drawer yielded the following result (Table 4):
With the equation being:

$$\text{Cognitive load} = 36.875 - 10.132\text{EFComponent}$$

For the tabs design pattern, we have the following regression result (Table 5):
The regression equation being:

$$\text{Cognitive load} = 0.859 + 0.015\text{EFComponent}$$

Table 3 Regression results for the scrolling menu pattern

R	R squared	Adjusted R square	Std. error of the estimate
0.629	0.396	0.386	11.33

Table 4 Regression results for the navigational drawer pattern

R	R squared	Adjusted R square	Std. error of the estimate
0.595	0.354	0.343	13.915

Table 5 Regression results for the tabs pattern

R	R squared	Adjusted R square	Std. error of the estimate
0.446	0.199	0.186	0.0299

Table 6 Regression results for the one-windows drilldown pattern

R	R squared	Adjusted R square	Std. error of the estimate
0.509	0.259	0.247	14.702

Table 7 Regression results for the two-panel select pattern

R	R squared	Adjusted R square	Std. error of the estimate
0.382	0.146	0.132	0.0884

Table 8 Regression results for the wizard pattern

R	R squared	Adjusted R square	Std. error of the estimate
0.559	0.312	0.301	0.8004

For the one-windows drilldown pattern, the regression results were (Table 6):
With the regression equation being:

$$Cognitive\,load = 31.515 - 8.562EFComponent$$

For the two-panel select, the regression results were (Table 7):
With the equation being:

$$Cognitive\,load = 0.399 + 0.036EFComponent$$

And finally, for the wizard design pattern, the regression results were (Table 8):
The regression equation being:

$$Cognitive\,load = 4.120 - 0.535EFComponent$$

The results of the classification trees show some relatively effective rulesets, obtaining some promising rules for selecting some design patterns.

The regression analysis showed in that in some cases, the executive function moderately influenced the cognitive load produced by the patterns, in some cases it was almost negligible.

6   Conclusions and Discussion

In this chapter we presented some interaction rules based on executive functions. Some of the tests measurements were significant in determining how the rules behaved. Although not all of the classification trees were significant. In most of the cases the accuracy of the trees was well below the ten percent mark. It was mostly in the cases where near the middle of the pack in terms of usability testing results of the patterns that the trees weren't a good fit, but as seen in the results, the trees of the top patterns and the worst performing ones were the more accurate ones.

In the repeated measure ANOVA it showed a statistically significant difference between means of the cognitive load produced by each pattern, even with the covariate of executive functions, although not as significant. Even then, it shows that there is some relationship between the executive functions of working memory and cognitive load, and the interface design pattern used.

The regression analysis is another story, some patterns showed very little influence from the executive function component, while a few showed moderate influence nearing a 50% accuracy rate, it could probably be factored to other user attributes that for this study could not be controlled.

There is much future work to be done, this study can serve as a base in order to continue with a more throughout experiment. Due to limitations of the available population, there were aspects that we could not properly control, such as IQ of the participants, and a minimal standard deviation of the age, since there is a very limited number of, in this case, children with autism in our community where we made the study. All of those factors could have greatly impacted the results, so the results obtained of this study can't be said to be a good standard, but can serve as a base for future work.

References

1. Bauer F, Bolliet L, Helms HJ (1968) Report of a conference sponsored by the NATO Science Committee. NATO Softw Eng Conf 1968:8
2. Jacko JA (2012) Human computer interaction handbook: fundamentals, evolving technologies, and emerging applications. CRC Press, Boca Raton
3. Card SK, Newell A, Moran TP (1983) The psychology of human-computer interaction. CRC Press, Boca Raton
4. Kim GJ (2015) Human-computer interaction: fundamentals and practice. CRC Press, Boca Raton
5. Chisholm W, Vanderheiden G, Jacobs I (2001) Web content accessibility guidelines 1.0. Interact ACM 8(4):35–54
6. Leidner DE, Jarvenpaa SL (1995) The use of information technology to enhance management school education: a theoretical view. MIS Q 19:265–291
7. Byrne MD (2001) ACT-R/PM and menu selection: applying a cognitive architecture to HCI. Int J Hum Comput Stud 55(1):41–84
8. Kieras D (1994) GOMS modeling of user interfaces using NGOMSL. In: Conference companion on human factors in computing systems. ACM, pp 371–372
9. Kieras DE, Meyer DE (1997) An overview of the EPIC architecture for cognition and performance with application to human-computer interaction. Hum Comput Interact 12(4): 391–438
10. Anderson JR (1996) ACT: a simple theory of complex cognition. Am Psychol 51(4):355
11. Byrne MD (2001) ACT-R/PM and menu selection: applying a cognitive architecture to HCI. Int J Hum Comput Stud 55(1):41–84
12. Schneider-Hufschmidt M, Malinowski U, Kuhme T (1993) State of the art in adaptive user interfaces. In: Adaptive user interfaces: principles and practice. Elsevier, p 3

13. Stephanidis C, Paramythis A, Sfyrakis M, Stergiou A, Maou N, Leventis A et al (1998) Adaptable and adaptive user interfaces for disabled users in the AVANTI project. In: Intelligence in services and networks: technology for ubiquitous telecom services, pp 153–166
14. Gajos K, Weld DS (2004) SUPPLE: automatically generating user interfaces. In: Proceedings of the 9th international conference on intelligent user interfaces. ACM, pp 93–100
15. Jurado MB, Rosselli M (2007) The elusive nature of executive functions: a review of our current understanding. Neuropsychol Rev 17(3):213–233
16. Delis DC (2012) Delis-rating of executive functions. Pearson, Bloomington
17. Miller EK, Cohen JD (2001) An integrative theory of prefrontal cortex function. Annu Rev Neurosci 24(1):167–202
18. Lezak MD (1995) Neuropsychological assessment, 3rd edn. Oxford University Press, New York, p 49
19. Pribram KH (1973) The primate frontal cortex—executive of the brain. Psychophysiology of the frontal lobes. Academic, New York, pp 293–314
20. Luria AR (1973) The working brain: an introduction to neuropsychology. Basic, New York
21. Hongwanishkul D, Happaney KR, Lee WS, Zelazo PD (2005) Assessment of hot and cool executive function in young children: age-related changes and individual differences. Dev. Neuropsychol 28(2):617–644
22. Godefroy O, Cabaret M, Petit-Chenal V, Pruvo JP, Rousseaux M (1999) Control functions of the frontal lobe: modularity of the central-supervisory system. Cortex 35:1–20
23. Dix A (2004) Human computer interaction. Pearson Education, Bloomington
24. Mayer RE, Moreno R (2003) Nine ways to reduce cognitive load in multimedia learning. Educ Psychol 38(1):43–52
25. Miyake A, Friedman NP, Emerson MJ, Witzki AH, Howerter A, Wager TD (2000) The unity and diversity of executive functions and their contributions to complex "frontal lobe" tasks: a latent variable analysis. Cogn Psychol 41(1):49–100
26. Sweller J (1988) Cognitive load during problem solving: effects on learning. Cogn Sci 12:257–285
27. Ayres P, Sweller J (2005) The split-attention principle in multimedia learning. In Mayer RE (ed) The Cambridge handbook of multimedia learning. Cambridge University Press, New York, pp 135–146
28. Chandler P, Sweller J (1992) The split-attention effect as a factor in the design of instruction. Br J Educ Psychol 62:233–246
29. Ayres PL (2001) Systematic mathematical errors and cognitive load. Contemp Educ Psychol 26:227–248
30. Bratfisch O, Borg G, Dornic S (1972) Perceived item-difficulty in three tests of intellectual performance capacity (Tech. Rep. No. 29). Institute of Applied Psychology, Stockholm
31. Paas FG (1992) Training strategies for attaining transfer of problem-solving skill in statistics: a cognitive-load approach. J Educ Psychol 84:429–434
32. Britton BK, Tesser A (1982) Effects of prior knowledge on use of cognitive capacity in three complex cognitive tasks. J Verbal Learn Verbal Behav 21:421–436
33. Van Gerven PWM, Paas F, van Merriënboer JJG, Schmidt HG (2004) Memory load and the cognitive pupillary response in aging. Psychophysiology 41:167–174
34. Paas F, Ayres P, Pachman M (2008) Assessment of cognitive load in multimedia learning: theory, methods and applications. In: Robinson DH, Schraw G (eds) Recent innovations in educational technology that facilitate student learning. Information Age Publishing, Charlotte, pp 11–35
35. Antonenko P, Paas F, Grabner R, van Gog T (2010) Using electronencephalography to measure cognitive load. Educ Psychol Rev 22:425–438
36. Van Gog T, Rikers RMJP, Ayres P (2008) Data collection and analysis: assessment of complex performance. In Spector JM, Merrill MD, van Merriënboer JJG, Driscoll MP (eds) Handbook of research on educational communications and technology. Routledge, London/England, pp 783–789

37. American Psychiatric Association (2013) Diagnostic and statistical manual of mental disorders (DSM-5). American Psychiatric Publishing, Arlington
38. Hill EL (2004) Executive dysfunction in autism. Trends Cogn Sci 8(1):26–32
39. Korkman M, Kirk U, Kemp S (2007) NEPSY-II: clinical and interpretive manual. The Psychological Corporation, San Antonio
40. Wechsler D (2003) Wechsler intelligence scale for children-WISC-IV. Psychological Corporation
41. Laurie-Rose C et al (2014) Measuring perceived mental workload in children. Am J Psychol 127(1):107–125

Integrating Learning Styles in an Adaptive Hypermedia System with Adaptive Resources

Carlos Hurtado, Guillermo Licea and Mario Garcia-Valdez

Abstract At present, e-Learning for distance education is increasing, but most of them do not take into account the individualities of the students, such as learning styles, offering the same content to all. This paper presents, an adaptive hypermedia web system based on learning styles; At the start of the session, the user performs the Felder and Soloman learning styles questionnaire to obtain information from the students. These results show learning objects (OA) for the computer programming subject to the students to analyze later the data of interaction of them with the system and thus determine if these objects apply to that style of learning, in case they solve them in many attempts, feedback is sent to the teacher to modify them based on their learning styles and exercises that are resolved in fewer efforts to improve the course, when the system collects more user interaction data will make better recommendations to new users using simple sequencing.

1 Introduction

Learning styles definition is how people can learn. For Dunn and Price [1], learning styles reflect "the way basic stimulus affect a person's ability to absorb and retain information." Learning styles are used in psychology and education to refer to the different ways people solve problems or in other words how they respond to stimulus or information. In the area of education can be defined as the way students interacts in classes or learning environments.

C. Hurtado (✉) · G. Licea
Facultad de Ciencias Quimicas e Ingenieria, Universidad Autonoma de Baja California,
Calzada Universidad, Mesa de Otay, 14418 Tijuana, Baja California, Mexico
e-mail: churtado@uabc.edu.mx

G. Licea
e-mail: glicea@uabc.edu.mx

M. Garcia-Valdez
Departamento de Sistemas y Computacion, Instituto Tecnologico de Tijuana,
Calzada Tecnologico S/N, Tomas Aquino, 22414 Tijuana, Baja California, Mexico
e-mail: mariog@tectijuana.mx

© Springer International Publishing AG, part of Springer Nature 2018
M. A. Sanchez et al. (eds.), *Computer Science and Engineering—Theory and Applications*, Studies in Systems, Decision and Control 143,
https://doi.org/10.1007/978-3-319-74060-7_3

In the traditional university educational model, the teacher imparts his class to the group based on his experiences; however, it is taught based on the teacher's style ignoring individual differences of students as their learning style, causing lack of understanding many times in classes, exercises and they can experience symptoms like frustration, despair and often do not approve subjects, even some dropout school.

Within subjects of engineering students one of the most complicated is computer programming, this because it includes a different way of thinking; consequently, the students have difficulties to describe and to apply this thought, to teach the programming language is complicated [2] there are many reports [2–6] on teaching of programming in Java.

The results said that there are problems, challenges or difficulties in teaching programming in Java. Phit-Huan [7] identifies different dimensions in complexity learning programming courses, with the goal of proposing alternatives to teach programming such as previous computer knowledge, experience in programming, difficulties while learning programming and factors leading to a poor performance in programming [7]. Students find it difficult to understand until they acquire a mental model of how programming works and how it is stored in memory and how it all fits together [8]. Nedzad and Yasmeen [2] indicate that there are some problems with Java such as inputs and outputs, reserved words, syntaxis and that there is not much help available.

Information technologies use increase in all areas of human life, and education is not the exemption, more and more universities use educational platforms to distribute learning material among their students, even many offers blended or online courses, this due to the advantages that educational systems have, one of the most important is that it can adapt to every student based on factors such as difficulty level or learning styles, that students can see the learning material at any time, and as many times as they want, they can get feedback from teachers and teachers can get feedback from students as well, or even the system can collect interaction data of user to create better learning material.

This paper is organized as follows. Section 2 presents state of the art. Section 3 explains the state of the art of the system and all its components. Section 4 shows and describes the architecture and parts to make learning activities recommendations. Section 5 describes the adaptive hypermedia systems and its interface. Section 6 presents the implementation and the users' interaction results, and in Sect. 7 we show the conclusion, interpretation of the experiment and future works.

2   State of the Art

There are several models of learning styles such as the Representation System (NLP Model), information processing (Honey and Mumford), bipolar category (Felder and Soloman) and thinking preferences (Ned Herman). Within these models exist several questionnaires such as [9–14].

2.1 Index of Learning Styles Questionnaire

Some models are used more in areas depending on the field of knowledge and the learning styles described, in engineering area the one with higher acceptance is the questionnaire Index of Learning Style of Felder and Soloman the result is according to four learning styles (Active-Reflective, Sensitive-Intuitive, Visual-Verbal, and Sequential-Global) explained below [11].

2.1.1 Active and Reflective Learners

Active learners tend to retain and understand information best by doing something active with it—discussing or applying it or explaining it to others. Reflective learners prefer to think about it quietly first.

"Let's try it out and see how it works" is an active learner's phrase; "Let's think it through first" is the reflective learner's response.

Active learners tend to like group work more than reflective learners, who prefer working alone.

Sitting through lectures without getting to do anything physical but take notes is hard for both learning types, but particularly hard for active learners.

2.1.2 Sensing and Intuitive Learners

Sensing learners tend to like learning facts, intuitive learners often prefer discovering possibilities and relationships.

Sensors often like solving problems by well-established methods and dislike complications and surprises; intuitors like innovation and dislike repetition. Sensors are more likely than intuitors to resent being tested on material that has not been explicitly covered in class.

Sensors tend to be patient with details and good at memorizing facts and doing hands-on (laboratory) work; intuitors may be better at grasping new concepts and are often more comfortable than sensors with abstractions and mathematical formulations.

Sensors tend to be more practical and careful than intuitors; intuitors tend to work faster and to be more innovative than sensors.

Sensors don't like courses that have no apparent connection to the real world; intuitors don't like "plug-and-chug" courses that involve a lot of memorization and routine calculations.

2.1.3 Visual and Verbal Learners

Visual learners remember best what they see—pictures, diagrams, flow charts, time lines, films, and demonstrations. Verbal learners get more out of words—written and spoken explanations. Everyone learns more when information is presented both visually and verbally.

2.1.4 Sequential and Global Learners

Sequential learners tend to gain understanding in linear steps, with each step following logically from the previous one. Global learners tend to learn in large jumps, absorbing material almost randomly without seeing connections, and then suddenly "getting it."

Sequential learners tend to follow logical stepwise paths in finding solutions; global learners may be able to solve complex problems quickly or put things together in novel ways once they have grasped the big picture, but they may have difficulty explaining how they did it.

2.2 Learning Systems

Currently, there are different types of online educational platforms, such as management learning system (MLS), adaptive hypermedia systems (AHS) and intelligent tutoring systems (ITS). These systems integrate hypermedia material which incorporates contents such as audio, videos, maps, images and text; users interact with this content.

One of the disadvantages of these systems is that the material presented to students is the same for all, another difficulty is that users should be filtering information and they can be confused navigating, they may browse between learning activities that are not useful to them, causing frustration or lack of interest to use the educational platform ignoring the user, their navigation options, and the subjects when offering the content.

As a solution to these problems, they create adaptive systems which individually adjust to specific characteristics of students, such as learning style, exercises levels of difficulty and suggestions of navigation among others.

According to Garcia-Barrios [15], an Adaptive System is one that "upon receiving a stimulus, alters something in such way that the result of the alteration corresponds to a suitable solution to satisfy some specific needs", the adaptive systems modify aspects of its structure, functionality or interface to favor a user need or group of users.

Within the adaptive systems, the most used are the adaptive hypermedia systems, this personalizes the presentation of resources and navigation interfaces for the user. According to Brusilovsky [16–18] adaptive hypermedia systems construct a model of goals, preferences, and knowledge of each user, and it uses this model during interaction with the user.

Benyon and Murray [19] identify three essential components of an adaptive hypermedia system: the user model that represents knowledge and preferences that "believes" the user has; the model of domain that describes the experience that is in the interaction, it is essential to know the desired communication between the System and the users, capture information generated by the interaction and define the inferences, adaptations, and evaluations that may occur.

Several studies on the adaptive hypermedia systems have reached the following conclusions

1. AHS positively affects students' performance [20–27].
2. AHS increases perceived learning outcomes [28, 29].
3. AHS facilitates learning processes [30–34].
4. Levels of satisfaction were higher [23, 28, 31, 35–37].
5. AHS are easy to use [23, 26, 33].
6. AHS were useful and helpful [24, 31, 33, 38].
7. Control of students in AHS is preferable [31, 39].
8. Most students accept the AHS recommendations [33, 40].
9. AHS had no significant effect on learning outcomes [28, 29, 39, 40].

2.3 Learning Objects

One of the essential things in hypermedia systems courses is educational resources; these are called learning objects (LO) the characteristics they are:

Reusable: they serve as a basis for the creation of other courses and others educational resources.
Must be durable: regardless of the updates made they must continue working.
Portable: they are moved and stored on various platforms, and their contents do not change.
Interoperable: can be used in different software and hardware platforms.
Accessible: with efficient location and retrieval using metadata.

Other characteristics of LO are that they allow exploring the course only in specific themes if required, just use the necessary activities, they are available at all times and will enable the creation of courses tailored to organizations or individuals [41]. Learning objects are a collection of one or more data, defined as an electronic representation of text, images, evaluations, or any format that can be represented by a web client.

2.4 Simple Sequencing

For metadata representation various learning objects standards are defined, being the specification of simple sequencing the most used, this proposed model is an extension of the Specification of Simple Sequence [42] one of its characteristics is

Fig. 1 Activity tree with
navigation rules

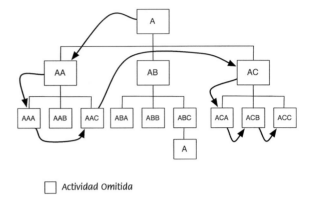

Actividad Omitida

that the main components of this standard are the Learning Activities and the Tree
of activities shown in Fig. 1, the learning activity is a pedagogically neutral unit of
instruction, knowledge, and evaluation, these activities can be nested sub-activities
and are organized in a tree-like structure called Tree of Activities, depending on the
application you can apply conceptual labels to learning activities.

There is a hierarchy of containers in the tree; depending on the application you
can apply conceptual labels in the tree, the leaf nodes are related to resources of
activities (the equivalent to Learning Objects). The advantages of these structures
are that they use the model of organization of folders or directories, because of this
the majority of users find this ordering natural.

Students navigate through the activity tree and receive activities one at a time
based on their learning style. To define a sequence in where students review
material the tree path made is in preorder. However, this order is modified by the
association of sequence rules to the nodes.

The rules are associated with the parent node and have scope only for the nodes
that belong to the cluster. An example of the navigation routes that can be seen in
Fig. 1 where the nodes labeled AAB and AB are omitted as an effect of rules
associated with these same nodes, when a node is ignored the nodes that it contains
too.

The rules can take into account the performance that a user had on his activities
tour, this way you can limit the access to specific resources. According to the
specification, the implemented System must have a follow-up of users' interaction
with resources; the rules consider this information, with this, navigation routes can
be defined [43].

3   Related Works

Adaptive hypermedia systems were developed to support learning styles their
adaptation, some of the following are:

IDEAL: Intelligent Distributed Environment for Active Learning (IDEAL) [44] is an Intelligent Adaptive System assisted by agents to support learning. The learning material adapts based on prior knowledge, learning styles, language and accessibility [44], this system provides characteristics of navigation and content adaptation. Students could redo the questionnaire or decide if the results will be used for all courses.

LSAS: Learning Style Adaptive System (LSAS) [21] uses the questionnaire [11]. The adaptation provided is by two different interfaces, one for sequential and other for global students. These elements give the students an overview of the subject and allow them to navigate in the course freely.

iWeaver: Its architecture [45] is based on Dunn and Dunn model of learning styles, incorporates several aspects of this model and its purpose is to have a balance between a student's cognitive load, an accessible browsing option, and learning content. When students use the System for the first time, they need to do the "Building Excellence Inventory" [46] to evaluate their learning style according to Dunn and Dunn's model. After each unit, the student gives feedback about effectiveness, progress, and satisfaction of the learning material.

INSPIRE: Intelligent System for Personalized Instruction in a Remote Environment (INSPIRE) [31] leaves students to select their learning objective and as consequence generates lessons that correspond to specific learning outcomes; accommodating the level of knowledge, progress, and students style of learning. Adaptation provided is regarding curricular sequencing, support for navigation, presentation, and curriculum sequencing; adaptive navigation is done based on students' objectives. The questionnaire developed by Honey and Mumford [47] is applied and filled out the first time they start a session.

AHA!: Adaptive Hypermedia for All (AHA!) [48, 49] allows authors to decide on the model of learning styles they want to implement in the course. They developed a tool for authors [50] as well as a generic adaptation language for learning styles called LAG-XLS [51]. This system allows three types of adaptation behavior: selection of elements to present, sort information and create different navigation routes. AHA! does not provide a questionnaire to identify learning styles. Instead, the user must register and manually choose their learning style.

4 Arquitecture

This section presents an educational adaptive hypermedia system [52], in which students must complete a couple of learning activities previously specified by the instructor. These learning activities may have information about student performance and learning styles to recommend a value based on student's profile. The instructors specify their recommendations using a Mamdani diffuse inference system with rules that have diffuse recommendation values as their consequent:

IF Visual IS Strong AND Verbal IS Mild THEN Recommended IS Low
IF Visual IS Mild AND Verbal IS Strong THEN Recommended IS High.

Learning activities are multimedia resources (i.e., video, text or audio) for example a learning activity presented in text format has a higher recommendation value for students with a strong verbal learning style. One way to add adaptive behavior to the System is changing parameters of membership functions in response to students' feedback.

In this algorithm, the same characteristics for recommendations are considered:

Learning styles: are determined by a test when users sign in for the first time, the questionnaire they resolve is the Felder-Soloman Index of Learning Styles for engineering students.

The recommendation algorithm considers three individual cases:

A new student added. To be precise, the collaborative filtering algorithm needs to know the students' previous preferences, based on the classification matrix values. When a new user is added to the system or has not made a certain number of classifications, there is not enough data to give accurate recommendations. If a student is "New" in the System, the instructor's gives a value based on students with similar learning style.

A new learning activity added. Each learning activity has standard metadata [53] indicating, among other things: target audiences, difficulty levels, formats, authors, and versions. This information can be used to make recommendations based on content when adding a new learning activity to the System. Also, considering instructor suggestions.

The architecture uses the object-oriented paradigm, with sharp and fuzzy attributes. Its components are domain model, prerequisite model, sequence model, context model, learning objects and basic resources.

The context model is composed of student model, session, workspace, and group. It represents the environment that surrounds the student, and interaction entities (in addition to the content of the learning objects): their partners, workspace, and technological support devices.

The user's model contains information from the user collected by the System, in the case of students it includes psychological factors which can be considered to make the customization as cognitive style, controls, and learning styles. Coffield [54] reports some of the leading objections we have about the estimation of learning style in particular:

- Measures of learning preferences based on subjective opinions that students make about themselves.
- Questionnaire items usually are ambiguous, do not consider the context or cultures where it may have another meaning.
- Learning style is only one of the factors that can influence learning, but it is not necessarily the most important.

In the group model, some techniques such as Differentiated Instruction (DI) or collaborative learning suggest the creation of user groups in a flexible way for instruction.

In the workspace, recorded information is to a physical environment in which the student is working.

The model of the domain elaborated by a group of instructors or based on some ontology, one of the disadvantages of this type of models is the subjectivity of the semantics of concepts; what is known as the problem of "landing symbols."

The prerequisite model, is modeled as a DAG (Directed Acyclic Graph) within the timeline; where circular nodes represent courses, or learning activities. The nodes are internally component of Activity Trees; there may be nodes not connected to the graph, in this case, the order specified is by its position in the timeline.

The sequence model uses simple sequencing specifications; learning activities have properties and fuzzy states.

4.1 System Architecture

Figure 2 shows the System architecture, the first time a user starts a session, the learning interface performs a learning styles questionnaire, in the end, results are stored in users profile database, activity tree uses profile information, later it uses feedback and effects of interaction with learning activities, to make recommendations. The teacher creates courses and learning activities based on data collected.

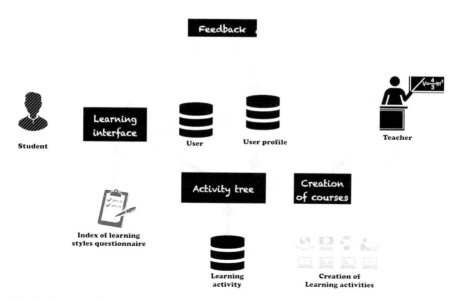

Fig. 2 System architecture

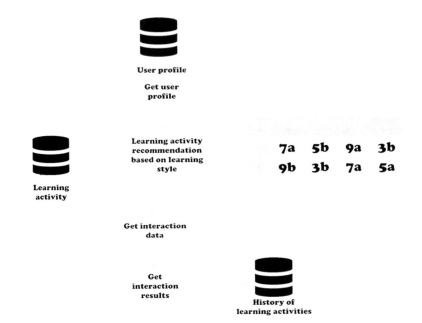

Fig. 3 User modeling diagram

4.2 User Modeling

When any user login the system, this reads information off his profile and makes a recommendation of learning activities based on his profile of learning styles, after using the system, interaction data and their results are stored in history of learning activities so based on their history of interaction with the material, recommendations are made to students with similar learning style. As shown in Fig. 3.

4.3 Recommendation of Learning Activities

To do feedback, the System reads a history of learning activities and user profile. Finally, it obtains learning activities based on the interaction of the user and the learning styles to make the recommendation (Fig. 4).

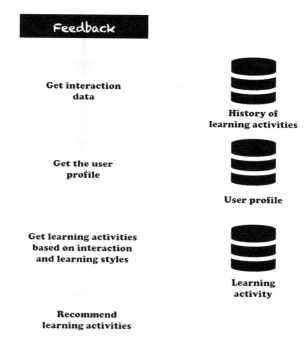

Fig. 4 Feedback and learning activity recommendation

5 The Adaptive Hypermedia System

This section shows the application developed and explains the interfaces.

5.1 Welcome Screen

Figure 5 shows the welcome screen, user must register the first time using the web page, registration can be through the system, Google or Facebook accounts.

5.2 Learning Styles Questionnaire

After starting session for the first time, the user must do the index of learning styles questionnaire of Felder and Soloman, this has forty-four multiple choice questions where you can answer one of two options based on what situation the person feels more identified with, after finishing, the system determines values for each learning style: active/reflexive, sensitive/intuitive, visual/verbal and sequential/global is the student (Fig. 6).

Fig. 5 Welcome and login screen

Fig. 6 Index of learning
styles questionnaire

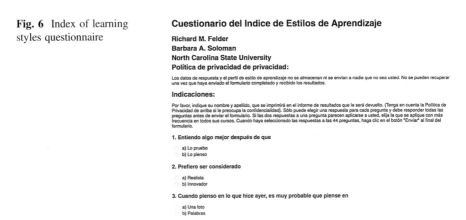

5.3 Creation of Learning Activities

Figure 7 shows the library of learning activities; here the teacher can create different material which can be texts, images, text with images, audios, videos, exams and programs, these can have labels, for example the learning style of that activity.

5.4 Creation of Courses

The course created is for the computer programming subject, for this, it is necessary to make a new course with its structure and learning activities, in the interface you can create containers used as units and inside them, there are activities added by the search submenu activity, which should be created previously in the library, Fig. 8.

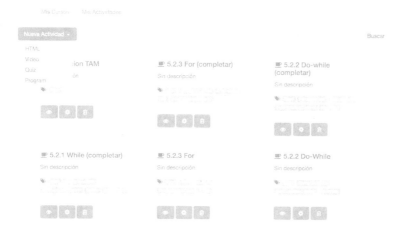

Fig. 7 Learning activities library

Fig. 8 Course creation and structure interface

5.5 Course Presentation Window

The course presentation is shown in Fig. 9. When taking the course, the structure is displayed depending on the rules that the teacher has registered when creating it, for example, some units or some subtopics can be hidden depending on the student progress or some tries.

5.6 Learning Activities

In Fig. 10 are some of the learning activities shown to the students, the first image is an example of a unit theoretical part, here you can add any multimedia content

Fig. 9 Computer programming welcome and course screen

Fig. 10 Learning activities

and the second image shows a programming exercise where users have to complete the code according to instructions, in case they commit some errors, the system sends feedback until students solve the problem correctly, the number of attempts can be configured by the teacher and data can be collected from interaction with the program.

6 Results

The System was implemented with groups of the informatics careers one of third semester and one of seventh semester in the Autonomous University of Baja California and a group of information and communications technologies

(ICT) career of seventh semester of the Technological Institute of Tijuana for a total of 60 students, both groups are currently enrolled in a computer programming course in Java programming language.

They were asked to respond the index of learning styles questionnaire, after that theory will be displayed and finally some evaluation programming exercises have to be solved, the unit to be evaluated was the number five regarding control and selection structures.

The results of the experiment are shown in Table 1 that most of the students are active, sensitives, visuals and sequential, it is noteworthy that visual learning style was the one offered a higher result within the students with 90% and the six students whose result was verbal it was low on the scale handled by the questionnaire, this means that they also have visual learning abilities. Another point of attention on the table is that students of seventh semester increased the percentage on visual and sequential style.

Table 2 shows the results of learning styles per career, in informatics, the predominant style is sensitive with 87.5%, and in ICT the visual style is with 96.43%.

After completing the learning styles questionnaire, the students performed twelve computer programming exercises on selection and control structures, divided into two sections. First, the student should study the theory of a sub-theme, for example, for the if statement they should program two exercises one to complete code and another to make a program from scratch, this is repeated until they finish the six subtopics.

The results of the attempts are presented in Table 3 this shows that the exercise to complete the else if statement was more difficult for the students to take an average of 3.32 attempts to solve it and the easiest to finish is the while structure taking an average of 1.34 attempts.

The right side of the table shows the results of attempts made by the students with visual learning style, they made more efforts of the exercise if else comp with an average of 3.22 attempts and the easiest was the while comp structure with 1.34 attempts.

Table 1 Total students learning styles results and 7th grade students results

Learning style	Quantity	Percentage (%)	Learning style	Quantity	Percentage (%)
Active	40	66.66	Active	28	66.66
Reflexive	20	33.33	Reflexive	14	33.33
Sensitive	50	83.33	Sensitive	35	83.33
Intuitive	10	16.66	Intuitive	7	16.66
Visual	54	90	Visual	39	92.85
Verbal	6	10	Verbal	3	7.14
Sequential	42	70	Sequential	31	73.80
Global	18	30	Global	11	26.19

Table 2 Learning styles results for students in informatics and information and communications technology

Learning style	Quantity	Percentage (%)	Learning style	Quantity	Percentage (%)
Active	20	62.5	Active	20	71.43
Reflexive	12	37.5	Reflexive	8	28.57
Sensitive	28	87.5	Sensitive	22	78.57
Intuitive	4	12.5	Intuitive	6	21.42
Visual	27	84.37	Visual	27	96.43
Verbal	5	15.62	Verbal	1	3.57
Sequential	21	65.62	Sequential	21	75
Global	11	34.37	Global	7	25

Table 3 Learning activities attempts and results of students with visual style

	Attempts	Average		Attempts	Average
If comp	113	1.88	If comp	103	1.9
If	146	2.5	If	121	2.3
If else comp	196	3.32	If else comp	171	3.22
If else	108	1.83	If else	96	1.81
Switch comp	116	1.97	Switch comp	108	2.04
Switch	132	2.24	Switch	125	2.36
While comp	75	1.34	While comp	67	1.34
While	99	1.77	While	83	1.66
Do while comp	88	1.57	Do while comp	79	1.58
Do while	93	1.66	Do while	82	1.64
For comp	165	3	For comp	156	3.18
For	93	2	For	84	2

7 Conclusion and Future Work

Based on the results the predominant learning styles in computer science subjects are active, sensitive, visual and sequential. The style of learning that counts with the highest percentage of students are visual and sensitive, that's why most of the learning objects must comply with visual characteristics and methods for solving problems must be well established which is of great help in making more and better learning objects.

The theoretical part has texts and examples of how to make the programs, so its possible to add more images for visual students, but in general, they did not make many attempts to solve the problems. However, we must analyze the examples that users complete in fewer attempts, to see how to improve the programs that require more efforts.

With the data collected, more learning objects will be created for the user to interact with different materials depending on their learning style, as more users join the System, this will make automatic recommendations to new users based on similar behaviors.

References

1. Dunn R, Dunn K, Prince G (1979) Learning style inventory (LSI) for students in grade 3–12. Price Syst 41
2. Nedzad M, Yasmeen H (2001) Challenges in teaching Java technology. Challenges Informing Clients A Transdiscipl Approach 365–371
3. Kolling M (1999) The problem of teaching object-oriented programming. J Object-Oriented Prog 11:8–15
4. Kolling M, Rosenberg J (2001) Guidelines for teaching object orientation with Java. In: 6th Conference on Information Technology in Computer Science Education (ITiCSE), pp 33–36
5. Madeen M, Chambers D (2002) Evaluation of students attitudes to learning the Java language. In: ACM International Conference, pp 125–130
6. Hristova M, Misra A, Rutter M (2003) Identifying and correction Java programming errors for introductory computer science students. SIGCSE Bull 35:19–23
7. Phit-Huan T, Choo-Yee T, Siew-Woei L (2009) Learning difficulties in programming courses: undergraduates' perspective and perception. In: International conference on computer technology and development, pp 42–46
8. Milne I, Rowe G (2002) Difficulties in learning and teaching programming—Views of students and tutors. Educ Inf Technol 7:55–66
9. Kagan J, Rosman BL, Day D, et al (1964) Information processing in the child: significance of analytic and reflective attitudes. Psychol Monogr 78
10. Kolb DA (1976) The learning style inventory: technical manual. McBer Co, Bost
11. Felder RM, Soloman BA (1988) Learning and teaching styles in engineering education. Engr Educ 78:674–681
12. Alonso C, Gallego D, Honey P (1999) Los estilos de aprendizaje
13. Mencke R, Hartman S (2000) learning style assessment
14. Renzulli J, Smith S, Rizza MG (2002) Learning styles inventory—version III. Creative Learning Press, CT
15. Garcia-Barrios VM, Mödristscher F, Gütl C (2005) Personalisation versus adaptation? A user-centred model approach and its application. En IKNOW 05 Graz 120–127
16. Brusilovsky P (1996) Methods and techniques of adaptive hypermedia. User Model User-adapt Interact 6:88–129
17. Brusilovsky P (2001) Adaptive hypermedia. User model user-adapt interact 11:87–110
18. Brusilovsky P, Nejdl W (2004) Adaptive hypermedia and adaptive web. CRC Press, USA
19. Benyon DR, Murray DM (1993) Adaptive systems; from intelligent tutoring to autonomous agents. Knowledge-based Syst 6(4):197–219
20. Alfonseca E, Carro RM, Martín E et al (2006) The impact of learning styles on student grouping for collaborative learning: a case study. User Model User-Adapt Interact 16: 377–401
21. Bajraktarevic N, Hall W, Fullick P (2003) Incorporating learning styles in hypermedia environment : empirical evaluation. In: Proceedings of the workshop on adaptive web-based systems, pp 41–52
22. Ford N, Chen SY (2000) Individual differences, hypermedia navigation, and learning: an empirical study. J Educ Multimed Hypermedia. https://doi.org/10.1.1.198.3170

23. Mampadi F, Chen SY, Ghinea G, Chen M-P (2011) Design of adaptive hypermedia learning systems: a cognitive style approach. Comput Educ. https://doi.org/10.1016/j.compedu.2010.11.018
24. Sangineto E, Capuano N, Gaeta M, Micarelli A (2008) Adaptive course generation through learning styles representation. Univers Access Inf Soc. https://doi.org/10.1007/s10209-007-0101-0
25. Taylor LL (2007) Adapting instruction based on learning styles for improved learning among rural community college students. University of South Alabama, Alabama
26. Triantafillou E, Pomportsis A, Demetriadis S (2003) The design and the formative evaluation of an adaptive educational system based on cognitive styles. Comput Educ 41(1):87–104
27. Triantafillou E, Pomportsis A, Demetriadis S, Georgiadou E (2004) The value of adaptivity based on cognitive style: an empirical study. Br J Educ Technol. https://doi.org/10.1111/j.1467-8535.2004.00371.x
28. Buch K, Sena C (2001) Accommodating diverse learning styles in the design and delivery of on-line learning experiences. J Work Learn 17(1):93–98
29. Siadaty M, Taghiyareh F (2007) PALS2: Pedagogically adaptive learning system based on learning styles. In: Seventh IEEE international conference on Advanced learning technologies, 2007. ICALT 2007
30. Graf S (2009) Adaptivity in learning management systems focussing on learning styles. In: Proceedings of the 2009. IEEE/WIC/ACM international joint conference on web intelligence and intelligent agent technology. https://doi.org/10.1109/WI-IAT.2009.271
31. Papanikolaou KA, Grigoriadou M, Kornilakis H, Magoulas GD (2003) Personalizing the interaction in a web-based educational hypermedia system: the case of INSPIRE. User Model User-Adapted Interact. https://doi.org/10.1023/A:1024746731130
32. Parvez SM, Blank GD (2007) A pedagogical framework to integrate learning style into intelligent tutoring systems. J Comput Sci Coll 22(3):183–189
33. Popescu E (2010) Adaptation provisioning with respect to learning styles in a Web-based educational system: an experimental study. J Comput Assist Learn. https://doi.org/10.1111/j.1365-2729.2010.00364.x
34. Tseng JCR, Chu HC, Hwang GJ, Tsai CC (2008) Development of an adaptive learning system with two sources of personalization information. Comput Educ. https://doi.org/10.1016/j.compedu.2007.08.002
35. Cabada RZ, Barron EML, Reyes Garcia CA (2011) EDUCA: a web 2.0 authoring tool for developing adaptive and intelligent tutoring systems using a Kohonen network. Expert Syst. https://doi.org/10.1016/j.eswa.2011.01.145
36. Essalmi F, Ben Ayed LJ, Jemni M et al (2010) A fully personalization strategy of E-learning scenarios. Comput Human Behav. https://doi.org/10.1016/j.chb.2009.12.010
37. Filippidis SK, Tsoukalas IA (2009) On the use of adaptive instructional images based on the sequential-global dimension of the Felder-Silverman learning style theory. Interact Learn Environ. https://doi.org/10.1080/10494820701869524
38. Schiaffino S, Garcia P, Amandi A (2008) eTeacher: providing personalized assistance to e-learning students. Comput Educ. https://doi.org/10.1016/j.compedu.2008.05.008
39. Brown E, Stewart C, Brailsford T (2006) Adapting for visual and verbal learning styles in AEH. In: Sixth international conference on advanced learning technologies
40. Wolf C (2007) Construction of an adaptive E-learning environment to address learning styles and an investigation of the effect of media choice. Sch Educ
41. Wiley DA (2000) Connecting learning objects to instructional design theory: a definition, a metaphor, and a taxonomy. Instr Use Learning Objects, Bloom Assoc Educ Commun Technol 3–23
42. Learning AD (2004) Sharable content object reference model, sequencing and navigation (SN)
43. Garcia-Valdez M, Licea G (2008) Aprendizaje colaborativo basado en recursos adaptativos
44. Shang Y, Shi H, Chen S-S (2001) An intelligent distributed environment for active learning. J Educ Resour Comput. https://doi.org/10.1145/384055.384059

45. Wolf C (2003) iWeaver : Towards 'Learning Style'-based e-learning in computer science education. Fifth Australas Comput Educ Conf (ACE 2003), 2003 Aust Comput Soc 2003 CRPIT ISBN 0-909925-98-4, pp 273–279
46. Rundle SM, Dunn R (2000) The guide to individual excellence: a self directed guide to learning and performance solutions
47. Honey P, Mumford A (1992) The manual of learning styles (3rd ed.). Peter Honey, Maidenhead
48. Bra P De, Calvi L (1998) AHA! an open adaptive hypermedia architecture. New Rev Hypermedia Multimed. https://doi.org/10.1080/13614569808914698
49. Stash N, Cristea A, De Bra P, Bra P De (2005) Adaptation to learning styles in E-learning : approach evaluation. Proc E-Learn 2006 Conf Honolulu, Hawaii
50. De Bra P, Aerts A, Rousseau B (2002) Concept relationship types for AHA ! 2. 0. Proc World Conf E-Learning Corp Gov Heal High Educ
51. Stash N, Cristea A, de Bra P (2005) Explicit intelligence in adaptive hypermedia: generic adaptation languages for learning preferences and styles. In: Proceedings of the international workshop on combining intelligent and adaptive hypermedia methods/techniques in web-based education, pp 75–84
52. García-Valdez M, Parra B (2009) A hybrid recommender system architecture for learning objects. Stud Comput Intell. https://doi.org/10.1007/978-3-642-04514-1_11
53. Memletics (2008) Learning styles inventory
54. Coffield F, Moseley D, Hall E, Ecclestone K (2004) Learning styles and pedagogy in post-16 learning A systematic and critical review. Learn Ski Res Cent. https://doi.org/10.1016/S0022-5371(81)90483-7

On Modeling Tacit Knowledge for Intelligent Systems

**Violeta Ocegueda-Miramontes, Antonio Rodríguez-Díaz,
Juan R. Castro, Mauricio A. Sanchez and Olivia Mendoza**

Abstract In an attempt to support efforts to narrow the gap between current Artificial Intelligence and actual intelligent human behavior, this paper addresses Tacit Knowledge. Tacit Knowledge is analyzed and separated into articulable and inarticulable for ease of scrutiny. Concepts and ideas are taken up from knowledge management literature aiming to understand the scope of knowledge. Among the bailed out concepts "particulars" and "concepts" stand out, and "preconcept" is suggested as an intermediate phase between the former two. These concepts are placed into mental processes of knowledge resulting in an alternative neurological model of knowledge acquisition. The model's target is to provide a picture as detailed as possible of the processes executed by the brain to make learning achievable. It encompasses from sensing the stimuli that is produced by the environment that are collected by sensory receptors to turn them into electrical impulses that are transmitted to the brain to climax with the emergence of concepts, from which increasingly complex knowledge is built. The model is then expanded to the social level.

1 Introduction

One of Artificial Intelligence's (AI) biggest challenges is to mimic human behavior, yet the goal is still far away. Therefore, a lot of effort has been devoted to find ways to create machines that can actually perform an "understanding" of concepts, or be able to develop skills, [1–10]. Although many relatively successful models have been proposed and used, they are very limited in practice [7]. For example, Artificial Neural Networks (ANN), are inspired by how the brain works, however they are implemented in fixed architectures, where neurons do not connect by

V. Ocegueda-Miramontes (✉) · A. Rodríguez-Díaz · J. R. Castro · M. A. Sanchez
O. Mendoza
Facultad de Ciencias Químicas e Ingeniería (FCQI), Universidad Autónoma
de Baja California, Tijuana, Baja California, Mexico
e-mail: vocegueda@uabc.edu.mx

© Springer International Publishing AG, part of Springer Nature 2018   69
M. A. Sanchez et al. (eds.), *Computer Science and Engineering—Theory
and Applications*, Studies in Systems, Decision and Control 143,
https://doi.org/10.1007/978-3-319-74060-7_4

themselves as they would in the brain. Besides, an ANN is a black box and carries on learning by trial and error. The reason it is a black box is that it deals with a kind of knowledge which is not easy to describe since it is hidden in the weights given to neuron's connectors. Same as the brain where it is still not known how, from neuronal activity, a mind emerges, which we consider it is the place were knowledge is constructed. The main problem is that only a small amount of this knowledge is known, so what happens with all other knowledge?

This other knowledge is known as "Tacit Knowledge" (TK). To address this TK there are seemingly two paths to follow: (1) open the "black box" and study whatever happens in there, which is something researchers will eventually be forced to do; or, (2) go back to nature and seek for inspiration for the sake of modeling the brain as close as possible to reality as understanding and available technology allows. In this study we propose to carry on the second path and forge a model closer to how the brain works (as far as we are aware of), that enables concept creation.

2 Tacit Knowledge

Tacit Knowledge (TK) was first coined by Polanyi, however it has been constantly redefined by researchers. Polanyi defines TK as "*unformulated knowledge, such as we have of something we are in the act of doing*" [11]. According to Howells, TK is "*non-codified, disembodied know-how*" [12]. For Stover, TK refers to "*the undocumented and unarticulated knowledge held by practitioners*", besides being "*intuitive and practice-based, which makes it both valuable and difficult to pass on to others*" [13]. Nonaka describes TK as being "*tied to the senses, tactile experiences, movement skills, intuition, unarticulated mental models, or implicit rules of thumb*" [14]. Collins perceives TK as knowledge "*which has not or cannot be made explicit*" [15].

According to Polanyi [16], for something to be considered knowledge it has to be linked to a tacit component. Consequently, TK can only be hold by the mind. Unfortunately, the mind is a very difficult concept to define because of its abstract nature, and there is little consensus on its meaning. Looking into dictionaries, there can be found different meanings, such as: (a) "*the same as understanding*" [17]; (b) "*the same as spirit, that is, the set of the superior functions of the soul, understanding and will*" [17]; (c) "*the collection of possible mental states and processes, whether they be affective, cognitive or volitional, of humans and other higher vertebrates*" [18]. Classical psychology proposes that the mind is divided into three parts: CsT (conscious mind), PCS (preconscious mind) and Ucs (unconscious mind). In consonance, the conscious mind is composed of "*those thoughts, feelings, motives, and actions of which we are phenomenally aware at the moment*" [19]. The preconscious mind consists of "*mental contents not currently in conscious awareness, but which were available to consciousness, and which could be brought into awareness under certain conditions*" [19]. Last, but not least, the unconscious mind embraces *"mental contents that are unavailable to consciousness-that could not enter awareness under any circumstances"* [19].

In one of his works, Kilstrom [19] takes up Freud's work in order to provide a historical review of the evolution of the concept of mind, which does by contrasting it to other researcher's understanding of the concept, such as: Pierre Janet, Claude Bernard, Morton Prince, Ernest R. Hilgard, among others. He subsequently redefines two of these concepts: the preconscious and the unconscious mind. Kilstrom addresses the preconscious mind as *"mental contents that could be accessible to conscious awareness, if conditions were right"* [19]. Besides, he conceptualizes the unconscious mind as *"those [mental contents] that are inaccessible to conscious awareness in principle, under any circumstances"* [19]. For the purpose of this work both definitions work perfectly fine, as we don't really perceive major differences between them. However, seeking to clarify what we mean with each of the concepts, we will stick to those defined by Freud as he contemplated the three definitions needed to continue on our work, and Kilstrom considers only two of them.

In this paper, TK is considered to be *"knowledge of which the individual is not conscious of possessing it"*, and so it can be said that it is related to the unconscious and the preconscious mind. On the other hand, EK resides in the conscious mind, which strongly depends on language. An appropriate definition for what we mean when referring to language is the one provided by Bunge, which states that it regards to *"signs system used to communicate and think"* [18]. We turned to this definition because the term "sign" encompasses a lot more things than mere words, such as symbols, images, gestures, sounds, etc.; which gives a better understanding of what we want to imply when alluding to EK. If both kinds of knowledge exist only within the mind, then our interpretation of whatever is in it, is not knowledge, but pointers to knowledge. Ribeiro himself points out that *"it only makes sense to say something is 'explicit' if we include it with the answer to the question: explicit for whom, when and in which social group?"* [20]. For example, if words were knowledge, then when a teacher spoke to his class, everyone in it should understand exactly the same. However, we do know that everyone understands things based on their context, and an individual's context is built up from his experiences in life. Therefore, it is impossible for two people to possess the exact same context. Thereupon it can be inferred that individuals understanding of things may differ from each other, even when referring to the exactly same thing (word, draw, etc.).

Another example is when two experts of an art are talking about a related topic and a novice is listening. Even when the three of them speak the same language, and the novice knows most of the words the experts are using, he is mostly unable to understand what they're saying. That's because, whilst what the experts are talking about is explicit for them, it is not for the novice. That is, the experts hold up TK related to the pointers they're using to communicate, which enables them to produce associated EK; the novice lacks correlated TK, and so cannot create the required EK to understand them. And that is because, as stated before, words (or any other form of expressing knowledge) are not knowledge themselves, but pointers to it. Taking this into account, it could be said that TK and EK existence began since the mind, as we conceive it, emerged.

2.1 Types of Tacit Knowledge

It is possible to analyze knowledge in two ways. First, as the continuum from TK to EK and vice versa. Second, as the dichotomy of know-how and know-what. Let's start by talking about know-how and know-what. Those two form a dichotomy, that is, knowledge either belongs to know-what or to know-how, but cannot belong to both or to a place in between. Know-what refers to facts. Whereas know-how relates to actions and practices; i.e., behavior. Wryly, despite their different nature, we claim that there cannot exist know-how without at least one know-what. Actually, we understand know-how as set of relations between know-whats. Still, to own a know-how is not to say that all know-what related to it is possessed. But, know-what's can eventually become known, and at some point those will be sorted, so that those collectively meet the know-how's goal. Know-what and know-how play along the tacit and explicit continuum. That is, they can either be tacit or explicit. On the other hand, TK and EK form a continuum in which the level of language and the level of consciousness can be observed. Which implies that TK, or at least part of it (this is clarified later on), can eventually become EK.

2.2 Components of Knowledge

We propose that TK and EK can be further understood in terms of so called "particulars", "pre-concepts", and "concepts". Following sections attempt to better explain these terms.

2.2.1 Particulars

Particulars is the most important term, because on its understanding relies the comprehension of pre-concepts and concepts. The term "particulars" has its origins in Polanyi's work [16]. Despite the fact that no precise definition is provided, he walks us through explanations to grasp its meaning. For a start, particulars do not hold a meaning by themselves. What encloses significance is the concept that they compose. Particulars cannot be reflected on because those are "*situated behind the barrier of language*" [11]. So even when individuals use them for reasoning, they can never be aware of them: "*I know the particulars of what I know only in an instrumental manner and am focally quite ignorant of them; so that I may say that I know these matters even though I cannot tell clearly, or hardly at all, what it is that I know*" [16].

For knowledge to be grasped, those particulars that support its understanding must have already been formed: "*Thus the meaning of a text resides in a focal comprehension of all the relevant instrumentally known particulars*" [16]. Which purport the idea that each individual relies on his already acquired particulars to provide information with a meaning. Knowledge, in all its forms, is raised on particulars: *"The arts of doing [know-how] and knowing [know-what], the valuation and the understanding of meanings, are thus seen to be only different aspects of the act of extending our person into the subsidiary awareness of particulars which compose a whole"* [16]. *"If a set of particulars [concept] which have subsided into our subsidiary awareness lapses altogether from our consciousness, we may end up by forgetting about them altogether and may lose sight of them beyond recall. In this sense they may have become unspecifiable"* [16]. I.e., when individuals learn to drive, at first they have to focus on every single step they have to follow, but when they finally assimilate the technique they suddenly begin to do it automatically, preconscious mind takes over, and so they are able to focus on other things while driving. They eventually get used to do it this way and forget how they do a lot of things (such as how to take a curve, how to speed up, when to start braking, etc.) despite always doing it. To conclude, we understand "particulars" as the basis of all knowledge, which lack significance by themselves, and can only hold a meaning when put together.

2.2.2 Pre-concepts

Basic pre-concepts result from massive connections between particulars inside the brain. Pre-concepts have different levels of complexity, which depend on the amount and kind of connections made. As stated above, pre-concepts may firstly originate from linking particulars, however, eventually pre-concepts begin to associate to other pre-concepts, giving rise to increasingly complex pre-concepts.

2.2.3 Concepts

Some researchers have taken on the task of providing the term "concept" with a definition. Next, some of them are cited: "a concept of x is a body of information about x that is stored in long-term memory and that is used by default in the processes underlying most, if not all, higher cognitive competences when they result in judgments about x" [21]; "that which allows people to have propositional attitudes (beliefs, desires, etc.) about the objects of their attitudes" [21]; "a concept is a temporary construction in working memory to represent a category" [22]; "a person's cognitive representation of a category on a particular occasion, regardless of its accuracy" [22]; "the basic timber of our mental lives" [23]; "concepts are mechanisms of detection, which allow us to track things, and which enable us to simulate them when they are not impinging upon our senses" [23].

From a computational point of view, these definitions do not serve our purposes. Following the argument stated above, we will consider concepts as an emergent property from a massive linkage of pre-concepts inside the brain, in the same way as particulars give rise to pre-concepts.

2.3   A Further Analysis of Knowledge

As already said, knowledge is associated to "parts" of the mind, so to speak. EK is related to the conscious mind. But, to better understand TK, it is crucial to make a distinction between its types. TK can be divided into two kinds: inarticulable and articulable TK. TK is linked to two "parts" of the mind: unconscious and preconscious. Thus, inarticulable TK is located in the unconscious mind; and articulable TK is found in the preconscious mind.

2.3.1   Inarticulable Tacit Knowledge

As mentioned above, inarticulable TK is enclosed in the unconscious mind and, by definition, could never be taken to consciousness. This type is made up of such primary forms of knowledge that there is merely no human language (at least not yet) that could display it. Therefore, it is not reflectible (actually, in principle it has no sense for the conscious mind) [16]. Ergo, it can never be articulated. Regarding this kind of knowledge, we suggest "particulars" (introduced by Polanyi [16]) and "pre-concepts" as its foundation. Instinct would be a good example of this type of knowledge.

2.3.2   Articulable Tacit Knowledge

Articulable TK resides in the preconscious mind, but, unlike inarticulable TK, it can be brought to consciousness. That is, it is amenable for articulation. And, although it is not yet encoded in any form of human language, it can eventually be expressed. This kind of knowledge is comprised of what we call "concepts". It should be made clear that these concepts are not yet consciously known. Albeit, we claim that a concept is the minimum piece of knowledge that can be uttered. Intuition works as an instance of this kind.

2.3.3   Explicit Knowledge

EK remains in the conscious mind. Due to its nature, it has to have a form of representation, which can range from words, sounds, images, gestures, movements, and so on. It is built from concepts of which individuals are consciously aware (in any of its representations). It should be made clear that, the fact that an individual cannot name something does not mean that he is not conscious of it, solely that he does not have words to express it. The conscious mind may have much more things than just words related to a concept, such as images, sounds, smells, texture, hardness, etc.

On the other hand, articulation of concepts is gradual (which, as stated before, refers to pointers to knowledge) and obviously depends on the handling of language the individual has. It usually starts from a very primitive representation (low level language) and it gradually becomes an increasingly complex one (high level language).

2.3.4 Relation Between Tacit and Explicit Knowledge

Inarticulable TK, articulable TK and EK form a continuum, or what is the same, the mind is a continuum. Thereby not meant that what is in the unconscious mind (inarticulable TK) can travel up to the conscious mind (EK), as to utter something, merely that any part of the mind may be affected by any other of its parts. Ergo, what happens in the unconscious mind (inarticulable TK) may influence the pre-conscious mind (articulable TK), which in turn may induce changes in the conscious mind (EK). It is worth noting that any interaction between the unconscious (inarticulable TK) and the conscious mind (EK) must always go through the pre-conscious mind (articulable TK). As a way to illustrate this point, it can be said that the unconscious mind (inarticulable TK) holds knowledge inherited from ancestors that could impact the concepts located in the conscious mind (EK).

The continuum proposed in this section contemplates: (a) two kinds of TK (inarticulable and articulable); (b) know-how and know-what; (c) knowledge relation to the mind; and, (d) knowledge relation to language. Figure 1 depicts the continuum suggested.

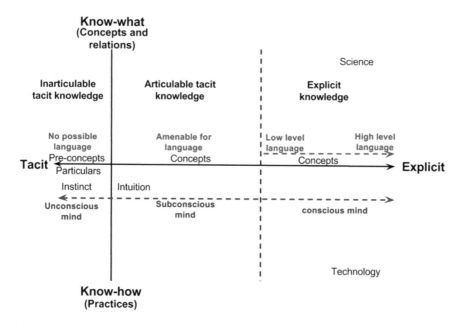

Fig. 1 Scheme of knowledge concept

2.3.5 Know-What and Know-How

Know-what refers to concepts and its relations. It can be both tacit and explicit. Therefore, it can take place in any part of the mind. Science is a good example of a pointer to it. On the other hand, know-how regards to practices individuals perform. And, alike know-what it can be held by any part of the mind. Technology is a good example of a pointer to it.

2.3.6 Synthesis of Knowledge

There may be concepts within an individual mind without it being related to a word. The individual can even be conscious of such concept and just does not know how to name it. I.e., babies know no words, so when a baby is hungry, for instance, all he does is cry. Every time he cries, his parents, among other things, try feeding him with a bottle. Sooner or later, the baby begins to associate the bottle with satisfying his hunger, and so, starts to identify the bottle. Thus, whenever he sees the bottle approaching he stops crying, because he knows his discomfort is reaching to an end. That is, the baby does know the concept, so much that he is able to recognize it and act accordingly. He knows its form, smell, taste, size, color, purpose, etc. He just does not know its name.

A concept is built up of many attributes, such as: form, color, size, smell, taste, sound, hardness, malleability, components (parts), purpose, context, meaning (relations to other concepts) and of course, words, inter alia. A concept is a set of relations, where words are just one of these many relations. It can be said that, the elements that produce a concept form part of all types of knowledge. That is to say, a concept is yield by particulars, which in turn form preconcepts, which jointly make up concepts, of which some may be part of articulable TK, and some may be components of EK. Going back to the baby example, it is possible to identify different manifestations of knowledge in it. (1) At least at first, when babies cry, they do it by instinct, they don't know why, they just do it. Instinct is inarticulable TK, is knowledge inherited from ancestors whose purpose is the preservation of the species. (2) Smell, taste, purpose, context, etc. are things babies know tacitly, that is, they have no representations of those, but could eventually have. That is articulable TK. (3) form, size, color, etc. are things babies are conscious of, that is, they have them as a mental representation of what a bottle is, and so, it is EK. Actually, in a baby it is possible to suppose the construction of a concept. When a baby is born it is said that he recognizes his mom by her smell. So he has the concept of mother, just a very basic one (where obviously no words are involved). After a few days of hearing her voice, he begins to identify it, so now he adds a voice to his concept. Later on, he starts to recognize his mom's face. Afterwards, he realizes that every time he cries, the one person that runs to his help is his mom, and so relates her to a purpose. Eventually, he is introduced to the word "mom" to refer to his mother, and so on. So children construct their concepts little by little, adding one thing at a time, and gradually forming increasingly complex concepts.

Some of the attributes of a concept, so to call them, are more complex to learn than others. One of the most difficult to get a hold of is context. It requires a lot of observation if done alone, a very accurate explanation (rules), or explanations based on cases. For example, when children are learning to talk, they have a lot of difficulties understanding the meanings and uses of some words. Such as the difference between the words "do" and "make". Some children may say something like "mom, I make my homework", to what the mother may reply "no, honey, it's not make, it's do, you do your homework". The child can get a little confused and ask the mother to explain when to use each one. Usually when an adult speaks a language, even if fluently, he is more likely to have forgotten most of the rules and just not be consciously aware of them (he tacitly knows them). Suppose that is the mom's case, and so she can only resort to explanations based on cases. Therefore, she teaches his child the difference of use between these two words through examples. Maybe, the child later turns to an English teacher, who surely would know the rules governing the use of both words (he explicitly knows them). Whatever the case, what the child is learning is context.

3   Tacit Knowledge's Acquisition Process

Our proposal is based on the assumption that when an individual observes a phenomenon several times, he begins to identify the things that characterize it. For each thing that he recognizes, connections between neurons are created within his brain. Those connections are called synapses. As already stated, particulars (inarticulable TK) can be explained as something that emerges from the interaction between neurons. Once particulars are created, those, somehow, begin to connect to each other creating what we call "pre-concepts" (inarticulable TK). Eventually, pre-concepts correlate with other pre-concepts resulting in what we defined as concepts (articulable TK). Alike, concepts relate to one another building increasingly complex concepts, and thus leading to the formation of rules.

The emergence of concepts from synapses can be thought of as follows:

(1) The environment is always sending stimuli (energy) to the senses [24], when stimuli reach the absolute threshold, other way it goes unnoticed [25–27], receptors for that kind of stimuli are activated [25].

    (a) However, some researchers argue that there are stimuli below the absolute threshold strong enough to activate receptors, just not strong enough for individuals "to be consciously aware of them" [25, 28]. Which means that there is consciously and non-consciously perception.

    (b) As all individuals differ from one another, it can be deduce that absolute thresholds for sensory organs are different for each single individual, e.g., an individual with hearing problems will have a higher absolute threshold for sounds than an individual with an average hearing. This means that the former individual will need stronger sounds to activate his sound receptors than the latter.

(2) Individuals perceive their surroundings through the senses, such as sight, hearing, taste, smell, and touch, to mention the primarily recognized.

  (a) Each sense is associated to a sensory organ [24].

  (b) All senses operate in a similar way, what makes them different is the kind of stimuli they detect and the specialized region of the brain they activate [26].

  (c) The sensory receptors (located in their respective sensory organ) gather the stimuli from the environment (light, sounds, flavors, odors, pain, and so on) and, through transduction, transform them into impulses (neural messages) [24–26, 28].

  (d) Impulses are sent to the specialized section of the brain that process that kind of information [24–26, 28]. Obviously, the environment does not send just one form of stimuli at a time, but all forms of stimuli all the time, so there are normally more than one sensory organ's receptors sending impulses to the brain at the same time.

  (e) Once impulses reach the brain, perception occurs. We can say it is "*the process by which the brain organizes and interprets sensory input [impulses]*" [28]. This led us to think that perception is the process through which:

    (i) Impulses are fired between neurons through synapses [26].

      (1) When impulses reach neurons, neurotransmitters are sent through synapses between them [29, 30].

      (2) Lasting divergence in sensory inputs, codified in the form of impulses, triggers changes in the neuronal network structure, that is, the way synapses are connected: "*Lasting alterations in sensory input trigger massive structural and functional adaptations in cortical networks*" [31]. Which means that, depending on impulses, synapses may be created or inhibited.

    (ii) The brain interprets impulses.

      (1) Individuals are born with a set of patterns of connections (synapses) between neurons inherited from their ancestors through DNA [32].

      (2) As individuals experience the world, more and more synapses are created/inhibited according to their experiences [33, 34]. This, in turn, leads to changes in individuals DNA, which will be passed down to their offsprings [35]. Besides, it may also cause changes in other types of inarticulable TK and inarticulable TK.

      (3) Synapses form patterns that are nothing but interpretations the brain gives to the sensory input (stimuli) provided by the sensory organs. Interpretations (patterns) can only be created based on

the synapses that already exist in the brain. If a synapse or set of synapses changes, then the pattern (interpretation) changes.

(4) The smallest unit of interpretation is not comprehensible at the conscious level, but mainly processed by the brain. This is what we call particulars (a term proposed by Polanyi [16]). The more synaptic patterns, the more particulars there are.

(5) Specific collection of synapses results into specific particulars.

(6) When synaptic patterns are attached to other synaptic patterns, it somehow leads to relations between particulars giving rise to what we call pre-concepts.

(7) Pre-concepts, although more complex than particulars, do not yet make sense and, as particulars, are also stored in the unconscious mind.

(8) A growing connection between synaptic patterns heads to associations between pre-concepts.

(9) Relations between pre-concepts eventually reach a point where qualitative changes occur emerging what we call concepts, which are entities that may be accessed by the conscious mind.

(10) For a concept to remain the same, its synaptic patterns must be kept unchanged, otherwise the concept faces changes.

(11) Relations between concepts give rise to increasingly complex concepts.

Although the main proposed process of concepts formation is depicted in the steps described above, it is worth mentioning some other suppositions we made regarding these points.

- As it can be deduced from point 1 (stimuli may activate receptors even when below the absolute threshold), there is information that the individual catches without him even noticing [25, 28]. That information (impulses) may induce changes in the synaptic patterns that could lead to modifications in concepts; hence to variations in the way individuals perceive and act. Leading individuals not to be clear about how they came to a conclusion.

- The preconscious mind is in charged of incorporating all concepts to make sense of them.

- When individuals are presented with a situation, the unconscious and/or the preconscious mind evaluates it and provides them with the knowledge necessary to deal with such situation. When the preconscious mind is the one making the decisions, and it deems indispensable, it sends the knowledge to the conscious mind. Otherwise, keeps it in the preconscious mind and just sends the judgment to the consciousness allowing individuals to come to a conclusion.

- When an individual encounters a situation that makes a great impact on him (e.g., something that threatens his life or security), even when never faced again, the synaptic patterns that are created are hardly destroyed due to an instinct of conservation all individuals possess (inarticulable TK/unconscious mind). This instinct of conservation allows individuals to detect dangerous situations, quicker when already experienced a similar one before, and act accordingly to ensure safety.

- The unconscious and preconscious minds can also be feeded by the consciousness, by consciously acquiring information through observation, practice, study, etc., and assimilating it.
- The preconscious mind shows itself in every single act, even when not realized.

4   Tacit Knowledge Acquisition Based on Rule Following

Once we understand how knowledge is acquired, it is mandatory to find ways to convey knowledge. There are researchers that believe that rule following may serve as a mean to transfer knowledge. Therefore, some of them have devoted time and effort to study the relationship between TK and rule following. Next, some of the most prominent are exposed. On one hand we have Wittgenstein, who envisages TK as something not only acquired through immersion in society, but also as something not consciously present in conduct. That is, the individual catches the rules and follows them, and that is how he displays his TK [36]. On the other hand, Ryle states that knowledge can be attained by explicitly knowing the rules and, sooner or later, assimilating them. The assimilation of rules is done by attempting to follow them, which creates particulars within his brain that enable him to tacitly recognize other rules. Over time, he comes to master the rules, which implies that his brain has created the particulars and relations needed to do it, which, at some point, endow him with the ability to play the game without even stressing out about the rules, he just happens to know them (tacitly). Besides, rules of the game can also be grasped just by observing them and tacitly follow them. Thereupon, to be able to follow the rules of a game is enough to say that someone possess that TK. However, if someone just knows the rules but cannot follow them, then he is said to not bear that tacit knowledge [37].

Furthermore, Polanyi asserts that to be able to follow a rule implies that there is related TK within the individual's mind, mainly because even when there are explicit rules, those would hardly express everything that involves performing them. When an individual follows rules without even noticing them that connotes the holding of TK: "the aim of a skillful performance is achieved by the observance of a set of rules which are not known as such to the person following them" [16]. Nontheless, he also clarifies that knowing the rules of an art does not imply TK in itself, those are maxims that seve as a guide, but knowing them means nothing unless those are incorporated into the practice of the art [16]. In other words, to know the rules of an art may entail tacit know-what, but does not presuppose tacit know-how (rule following). Further, observation and imitation of an expert (rule following) enables TK transference by allowing recognition of the rules of the art, even those hidden to the expert himself [16].

Another point of view regarding rule following is the presented by Bourdieu, who understands TK as "knowing how" (rule following) instead of "knowing that". Bourdieu depicts tacit knowledge as something "which exists in a practical state in an agent's practice and not in their consciousness or rather in their discourse" [36]. Bourdieu introduces the concept of "habitus", which represents the process through which TK is assimilated. That is, according to Bourdieu rule following is the only road to TK's acquisition [36]. In short, to know something tacitly is largely demonstrated by the ability to follow the rules that govern it.

4.1 Proposed Interpretation of Rule Following

Wittgenstein proposes a set of interesting ideas in this regard, some of them are that "rule following" [38]:

- It's a social product.
- The expert decides when the apprentice has learned the rule.
- There's a reaction to a sign.
- Rule is equal to institution or custom.
- It's a regularity.

Another important contribution to this field is the one made by Bourdieu, who coins the term "habitus", which encompasses "rule following".

Summarizing those researchers, "rule following" involves know-how and know-what. Where know-how is related to practices, and know-what is a collection of concepts and their relations. Practice theory happens to be in charge of research regarding practices, and concepts and their relations, which in turn, jointly, constitute culture. Know-how is embodied in techniques, whilst know-what is addressed by epistemology. And science uses know-how to accomplish goals, and has its results in the form of know-whats. Furthermore, we strongly believe that "rule following" enables TK's acquisition and conveyance. To illustrate this point we will resort to an example. When a father is trying to teach his child how to ride a bicycle, he would normally disclose a set of explicit rules to his child, such as:

- Place yourself in the middle of the bicycle.
- Set a foot on a pedal.
- Sit on the seat.
- Move onto the seat.
- Place your second foot on the remaining pedal.
- Begin pedaling.

Yet, even when the father has told his child the rules to ride the bicycle, the child is more likely not to get it right the first time he tries, neither the second one, and most probably not the third one either. The child would surely ask how not to fall, to what the parent may answer with something like "balance yourself", which, by the way, will not help much. Anyway, the child will have to follow the rules his father told him, until he eventually masters riding a bicycle.

In conclusion, we understand "rule following" as the ability to incorporate a set of rules of an art to personal behavior. "Rule following" is characterized by being done to achieve goals, which are hold by individuals and/or societies who establish them to their best advantage.

4.2   From Individual Tacit Knowledge to Collective (Social) Tacit Knowledge

After analyzing the foregoing, we have come to notice that "rule following" occurs at two levels: individually and collectively. Let's begin by the individual level. When a child is born, if not all, most of his knowledge is inarticulable TK (from DNA) and so, anything he does, he does it instinctively. When newborns are taken to their mothers for the first time in order to feed them, some of them don't eat, they don't even try to suck. This is because they still do not realize that they had already been separated from their mothers, and expect things to be the same way they used to be: to be fed by the umbilical cord, without them having to do anything. However, after a while that the baby is hungry and is not fed, by survival instinct, he begins to cry expressing his discomfort. Once again is brought to his mother, who places him beside her breasts, and then, instinctively, the baby begins to suck. In both cases, the baby is following rules not consciously, but unconsciously known, dictated by the knowledge his DNA has provided him with. If he had not possessed that knowledge, he would surely die. Plain and simple, there is just no way to teach a newborn how to suck his mother's breast and explain to him that, that is the new way he is going to be fed. His parents, no matter how much they love him, solely do not have a way to communicate with him. The newborn has to know how to do it in order to survive (and that is why it is in the DNA). Let's now assume some months go by, babies don't like being alone, instead they love attention. When a baby is alone he begins to cry, and the mother immediately goes running. After a while the baby catches that whenever he cries his mother comes to him (discovers a rule). Then, whenever he is alone, he starts pretending to cry in order to get his mother's attention (follows the rule).

Now let's take a look at the collective level. There was a time when humans cohabited with their predators, which made them, piecemeal, develop skills that helped them stay alive, which were part of the evolution that allowed the species' perpetuation. One of these skills is face recognition. At that time, recognizing a predator's face was the difference between life and death. The faster an individual recognized a face, in case it was a predator's, the sooner he could run to safety. It was such a transcendental skill for human survival, that it joined the DNA and, since then, has been inherited from generation to generation to the present day. Nowadays, individuals are predisposed to recognize faces wherever they look at (clouds, darkness, etc.). They don't even have to try; they just follow the rule that is codified in their DNA. Another example of collective (but articulable) TK has to do with the rules individuals follow within society, even though they are seldom mentioned. I.e., the way an individual behaves with his family, friends, teachers, bosses, etc. While it is true that children are introduced to some of society's rules by their mothers, it is also true they learn some others by themselves, and they eventually start to intuit how to behave in other situations. For instance, their mothers teach them how to behave towards adults, such as teachers, relatives, etc. Notwithstanding, they learn by themselves, with few exceptions, how to behave in front of friends. So, in childhood they pick up those rules, and assimilate them into their articulable TK. As they grow up, and based on what they have learned so far, they begin to deduce how to behave in completely new situations, even when no one has ever told them how to do it, such as with bosses, policemen, etc. They just follow the rules in their mind.

Now let's take a look at how the individual level and the collective level are intrinsically related.

Neurons (individual level: IL) versus individuals (collective level: LV) Neurons are the basic units of what constitutes the brain, from which the mind arises, which is what make individuals be who they are. Individuals are the basic units of what composes a society, and who make it have its specific characteristics.

Synapses (IL) versus relationships (CL) Synapses, are relations between neurons, which are channels for communication between neurons. Those can also be seen as physical canals that establish everything needed for communication to occur. Relationships are abstract connections between individuals. Those may also be described as abstract entities that dictate protocols of communication between individuals (father-son, wife-husband, teacher-student, etc.).

Neurotransmitters (IL) versus alphabet, sounds, points, etc. (CL) Neurotransmitters are chemical substances, on which neurons rely to communicate to each other. Alphabet is a set of letters used in writing system. Sounds are vibrations that travel through the air or another medium and can be heard when they reach a person's or animal's ear. Both, alphabet and sounds, among other things, are used for communication between individuals.

Set of ordered neurotransmitters (IL) versus Message (CL) A set of neurotransmitters ordered in a singular way convey a specific message between neurons. Sets of letters or sounds ordered in a concrete way convey a specific message between individuals.

Rules for interpreting sets of neurotransmitters (IL) versus human language (CL) There must be rules between neurons that enables them to decipher the messages they exchange. Human language rules provides individuals with the information required to interpret messages.

Particular (inarticulable TK) (IL) versus DNA (CL) Particulars are the basis of all the knowledge an individual holds, it can never reach consciousness, but manifests itself in every single act. DNA is "*the material in which our hereditary information is stored*" [39], it cannot, under any condition, get to consciousness, yet it predisposes physical appearance, diseases, personality, skills, and so on.

Preconcept (inarticulable TK) (IL) versus Habitus (CL) Preconcepts are what result from connecting a set of particulars, but still they don't hold a meaning. Habitus are the rules on which an individual builds his behavior. However, a single individual's behavior does not determine anything in a society.

Concept (articulable TK) (IL) versus unspoken social rules (culture) (CL) Concepts are knowledge of which individuals are not conscious of, anyhow, those govern the thinking and (unconscious) action of an individual. Unspoken social rules are a set of recurring habitus among members of a society which are not specified anywhere, but followed by everyone or at least most of them.

Explicit knowledge (IL) versus science, law, etc. (CL) EK is knowledge that resides within individuals' conscious mind; it regulates the way in which they express themselves and act. Science, law, etc. are written rules that regiment societies' modus operandi.

5 A Proposal for Artificial Intelligence and Multiagent Systems

Regarding ANNs, our proposal suggests that neural architectures must not be static; hence a new design is required to enable neurons to connect in order to generate particulars. But then these particulars should form connections among themselves to produce preconcepts that conform inarticulable TK. Finally preconcepts connect to engender concepts, which are articulable in principle. This means that there should be two layers of connections, a layer where neurons connect directly with one another, and a second layer where sets of wired neurons connect to other sets of wired neurons.

Besides, following Table 1, it is critical to recognize how repetitive actions may spread through agent's societies. That is, we must identify how relations at the social level correspond to synapsis at the individual level. If the individual's

Table 1 Summarizes the individual-collective relation proposed in this section

Individual level	Collective (social) level
Neuron	Individual
Synapses	Relationships
Neurotransmitters	Alphabet, sounds, etc.
Set of ordered neurotransmitters	Message (set of ordered letters, sounds, etc.)
Rules for interpreting sets of neurotransmitters (brain language)	Human language rules (grammar)
Particular (inarticulable TK)	DNA
Pre-concept (inarticulable TK)	Habitus
Concept (articulable TK)	Unspoken social rules (culture)
EK	Science, law, etc. (culture)

architecture advises that connections are dynamic, in the social level it happens naturally due to agents' interaction. But when a recurring relation between agents' takes place, this equals to a synapsis. In addition, the accumulation of relations must also be done through layers as in the individual level, so it allows habitus detection, in such a way that the accumulation of habitus could be considered as cultured among a group of agents.

As architectures are equivalent, both in individual and in agents, this allows the model to be scalable. For example, by using the same architecture, we could consider a whole society as an agent and the relations amid societies as commensurate to synaptic connections.

6   Conclusions

Artificial Intelligence has been struggling to generate mechanisms that simulate intelligent behavior. We suggest that TK and its parts may improve future attempts to model human behavior. Therefore, in this paper we analyze and come to a synthesis of knowledge, in which knowledge consists of EK and TK, the latter is divided into articulable TK and inarticulable TK. As is implied by their names, the former can be accessed by the conscious mind and so it can be uttered; while the second one is knowledge that resides in the subconscious, ergo in principle it cannot be accessed by the conscious mind through the introspection process. However, this kind of knowledge can be extracted from analysing social behavior through computational techniques for pattern recognition; for example, data mining techniques. We propose that knowledge has three phases: (1) particulars, that is knowledge's basic unit, which dwells in the subconscious mind and belongs to inarticulable TK; (2) preconcepts, that are set of related particulars that also inhabit the subconscious mind as inarticulable TK; (3) concepts, that are collections of interconnected preconcepts, which at some point take a "quantum leap" from highly complex preconcepts to concepts. It would be interesting to analyze this "quantum leap" under the approach of granularity theories to determine when a preconcept becomes a concept. Concepts, as most of the knowledge reside mainly in the subconscious mind but, unlike particulars and preconcepts, those can reach the conscious mind. That is, concepts normally exist as articulable TK, but when the mind is focused on them they change into EK.

After explaining types of knowledge, its components and their relationships, we move onto a neurological elucidation of knowledge acquisition process. In the model we describe how knowledge is built from stimuli captured by sensory receptors. The model describes that stimuli is transformed into electrical impulses and transmitted to the part of the brain in charge of processing it, sent to neurons to alter synapses (either it be by reinforcing existing synapsis or creating new ones). A minimum amount of synapsis gives rise to particulars, synapsis between particulars originates preconcepts, and an unknown number of synapsis between preconcepts make concepts emerge.

References

1. Ertel W (2017) Introduction to artificial intelligence, 2nd ed. https://doi.org/10.1007/978-3-319-58487-4
2. Jackson PC (1985) Introduction to artificial intelligence. Dover Publications
3. Akerkar R (2014) Introduction to artificial intelligence. Prentice-Hall of India, Delhi
4. Flasiński M (2016) Introduction to Artificial Intelligence, 1st ed. https://doi.org/10.1007/978-3-319-40022-8
5. Bottou L (2014) From machine learning to machine reasoning. Mach Learn 94:133–149. https://doi.org/10.1007/s10994-013-5335-x
6. Jordan MI, Mitchell TM (2015) Machine learning: trends, perspectives, and prospects. Science 349:255–260. https://doi.org/10.1126/science.aaa8415
7. Bordes A, Glorot X, Weston J, Bengio Y (2014) A semantic matching energy function for learning with multi-relational data. Mach Learn 94:233–259. https://doi.org/10.1007/s10994-013-5363-6
8. Xuan J, Lu J, Zhang G et al (2017) A Bayesian nonparametric model for multi-label learning. Mach Learn 106:1787–1815. https://doi.org/10.1007/s10994-017-5638-4
9. Weghenkel B, Fischer A, Wiskott L (2017) Graph-based predictable feature analysis. Mach Learn 106:1359–1380. https://doi.org/10.1007/s10994-017-5632-x
10. Zaidi NA, Webb GI, Carman MJ et al (2017) Efficient parameter learning of Bayesian network classifiers. Mach Learn 106:1289–1329. https://doi.org/10.1007/s10994-016-5619-z
11. Polanyi M (1959) The study of man. The University of Chicago Press, Chicago
12. Howells JRL (1996) Tacit knowledge, innovation and technology transfer. Technol Anal Strateg Manag 8
13. Stover M (2004) Making tacit knowledge explicit: the ready reference database as codified knowledge. Ref Serv Rev 32:164–173
14. Nonaka I, Von Krogh G (2009) Perspective—Tacit knowledge and knowledge conversion: controversy and advancement in organizational knowledge creation theory. Organ Sci 20:635–652
15. Collins H (2010) Tacit and explicit knowledge. University of Chicago Press, Chicago
16. Polanyi M (1962) Personal knowledge: towards a post-critical philosophy. The University of Chicago Press, New York
17. Abbagnano N (2004) Diccionario de Filosofía. Fondo de Cultura Económica, México, D.F
18. Bunge MA (2007) Diccionario de filosofía. Siglo XXI: 221
19. Kihlstrom JF (1999) The psychological unconscious. Handb Pers Theory Res 2:424–442
20. Ribeiro R (2013) Tacit knowledge management. Phenomenol Cogn Sci 1–30
21. Machery E (2010) Precis of doing without concepts. Behav Brain Sci 33:195–206
22. Barsalou LW (1993) Flexibility, structure, and linguistic vagary in concepts: manifestations of a compositional system of perceptual symbols. Theor Mem 1:29-31
23. Prinz JJ (2004) Furnishing the mind: concepts and their perceptual basis. MIT Press, Cambridge
24. Henshaw JM (2012) A tour of the senses: how your brain interprets the world. JHU Press, Baltimore
25. Ciccarelli SK, White JN (2014) Psychology, 4th edn. Pearson, London
26. Goldstein E (2013) Sensation and perception. Cengage Learning, Boston
27. Zimbardo PG, Johnson RL, Hamilton VM (2012) Psychology: core concepts, 7th edn. Pearson, London
28. Myers DG (2014) Exploring psychology. Worth Publishers, Basingstoke
29. The Society for Neuroscience (2012) Brain facts: a primer on the brain and nervous system. Soc Neurosci
30. Wickens A (2009) Introduction to biopsychology. Pearson Education, London

31. Butz M, van Ooyen A (2013) A simple rule for dendritic spine and axonal bouton formation can account for cortical reorganization after focal retinal lesions. PLoS Comput Biol 9: e1003259

32. Dias BG, Ressler KJ (2014) Parental olfactory experience influences behavior and neural structure in subsequent generations. Nat Neurosci 17:89–96

33. Cheetham CEJ, Barnes SJ, Albieri G et al (2014) Pansynaptic enlargement at adult cortical connections strengthened by experience. Cereb Cortex 24:521–531

34. Barnes SJ, Finnerty GT (2010) Sensory experience and cortical rewiring. Neurosci 16: 186–198

35. Yu H, Su Y, Shin J et al (2015) Tet3 regulates synaptic transmission and homeostatic plasticity via DNA oxidation and repair. Nat Neurosci 18:836–843

36. Gerrans P (2005) Tacit knowledge, rule following and Pierre Bourdieu's philosophy of social science. Anthropol Theory 5:53–74

37. Ryle G (1949) The concept of mind. University of Chicago Press, Chicago

38. Wittgenstein L, Suarez AG, Moulines CU (1988) Investigaciones filosóficas. Crítica, Barcelona

39. Brandenberg O, Dhlamini Z, Edema R et al (2011) Module A—Introduction to molecular biology and genetic engineering. Biosafety Resource Book. Food and Agriculture Organization of the United Nations

Influence of the Betweenness Centrality to Characterize the Behavior of Communication in a Group

K. Raya-Díaz, C. Gaxiola-Pacheco, Manuel Castañón-Puga,
L. E. Palafox and R. Rosales Cisneros

Abstract The behavior of the distribution of a rumor must emerge according to the relations between the individuals. Taking as a reference that human society creates links of friendship through random encounters and conscious decisions, therefore, a rumor can be spread considering the degree of grouping that individuals have, also their location in the network and if they decide to cooperate or not. Considering the analysis of the topology that interconnects a set of individuals, relationships are detected between them that allow recognizing their centrality of degree, betweenness, and closeness. For a rumor to spread it requires that the individual has an incentive by which he decides to cooperate or not in the distribution of it. In this chapter, we propose an agent-based model that allows the identification of the central measures of each of the individuals that integrate a group which has a topology based on the Barbell's graph.

1 Introduction

Networks have been used to illustrate a system as a set of nodes joined by links, these nodes could represent persons and the links their relations. In [1] Estrada define a network or graph as a diagrammatic representation of a system. Some networks are defined as the pair $G = (V, E)$, where V is a finite set of nodes and E are the edges this representation is known as simple network, but there are other kind called weighted network defined by $G = (V, W, f)$ where V are the nodes, W are the edges or links that associates the nodes with an specific weight, and f is the mapping which associates some elements of E to a pair elements of V [1, 2]. This topology information of a network help us to describe it behavior, and interactions in micro and macro level.

K. Raya-Díaz (✉) · C. Gaxiola-Pacheco · M. Castañón-Puga · L. E. Palafox
R. Rosales Cisneros
Facultad de Ciencias Químicas e Ingeniería, Universidad Autónoma de Baja California,
14418 Calzada Universidad, 22390 Tijuana, Baja California, Mexico
e-mail: kraya@uabc.edu.mx

© Springer International Publishing AG, part of Springer Nature 2018
M. A. Sanchez et al. (eds.), *Computer Science and Engineering—Theory and Applications*, Studies in Systems, Decision and Control 143,
https://doi.org/10.1007/978-3-319-74060-7_5

2 Network Models

When is necessary express a model of a network an adjacency matrix, is way to do it. Adjacency matrix A_{ij} allow us identify the relations of the elements. Figure 1 shows an example of a random network and Table 1 its adjacency matrix Aij, where the edge take the value 1 when vertex v_i is connected to v_j and 0 otherwise [3].

An adjacency matrix is used to determine the path length and algebraic connectivity of the nodes. Observing Table 1 is easy to find the most interconnected nodes, which are the sum of one's in each row this measure is known as degree k [4].

Other classification of the nodes that is calculate by adjacency matrix is detect how many hubs are in the network. A hub refers to a node with several links that greatly exceeds the average this concept was introduce by Barabási and is frequently finding in a scale-free networks [5].

Figure 2 illustrate a scale-free network is a network whose degree distribution follows a power law [5], this law indicate that the probability of a node to create a

Fig. 1 Illustrate a random network

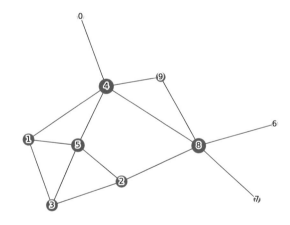

Table 1 Adjacency matrix of random network of Fig. 1

Nodes	0	1	2	3	4	5	6	7	8	9
0	0	0	0	0	1	0	0	0	0	0
1	0	0	0	1	1	1	0	0	0	0
2	0	0	0	1	0	1	0	0	1	0
3	0	1	1	0	0	1	0	0	0	0
4	1	1	0	0	0	1	0	0	1	1
5	0	1	1	1	1	0	0	0	0	0
6	0	0	0	0	0	0	0	0	1	0
7	0	0	0	0	0	0	0	0	1	0
8	0	0	0	0	1	0	1	1	0	1
9	0	0	0	0	1	0	0	0	1	0

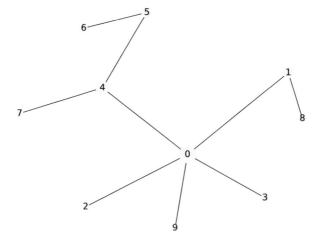

Fig. 2 Scale-free network example follow power law distribution

Fig. 3 Histogram of degree distribution in a random network

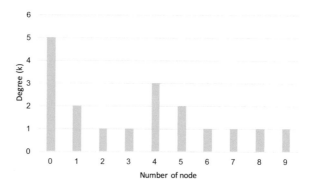

link with other node depends of the degree, in other words a node will prefer attach itself to a hub node than a peripheral node. In Fig. 3 shows the histogram with the degree distribution of each node of the network.

Power law is shown in Eq. 1 where *P(k)* is a fraction of nodes in the network with k links to other nodes and γ is a parameter whose values are in the range $2 < \gamma < 3$ typically.

$$P(k) \sim k^{-\gamma}$$

In Fig. 3 shown the degree distribution in which node cero is the preferred node in complex theory is called hub node. Barabási in [5] found that hubs radically shrink the distance between the nodes in scale-free networks.

After the analyzes of both kind of networks (random and scale-free) is time to introduce other graphs that reflects the topological structure in a network.

3 Barbell Graph

Barbell is a graph composed of two complete graphs with *m1* vertices, and one path that connect both graphs containing *m2* vertices [6]. Where the total number of vertices is equal to $n = 2m1 + m2$, and ($m1 \geq 2$, $m2 \geq 0$). The representation of Barbell graph is shown in Fig. 4, where we can see two bells connected by a path.

The adjacency matrix that represents the Barbell graph of Fig. 4, is illustrate in Table 2, where the number of vertices are 15 and the number of edges are 34.

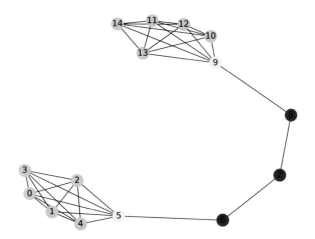

Fig. 4 Example of Barbell graph with m1 = 6 and m2 = 3

Table 2 Adjacency matrix of Barbell graph of Fig. 4

Nodes	0	1	2	3	4	5	6	7	8	9	10	11	12	13	14
0	0	1	1	1	1	1	0	0	0	0	0	0	0	0	0
1	1	0	1	1	1	1	0	0	0	0	0	0	0	0	0
2	1	1	0	1	1	1	0	0	0	0	0	0	0	0	0
3	1	1	1	0	1	1	0	0	0	0	0	0	0	0	0
4	1	1	1	1	0	1	0	0	0	0	0	0	0	0	0
5	1	1	1	1	1	0	1	0	0	0	0	0	0	0	0
6	0	0	0	0	0	1	0	1	0	0	0	0	0	0	0
7	0	0	0	0	0	0	1	0	1	0	0	0	0	0	0
8	0	0	0	0	0	0	0	1	0	1	0	0	0	0	0
9	0	0	0	0	0	0	0	0	1	0	1	1	1	1	1
10	0	0	0	0	0	0	0	0	0	1	0	1	1	1	1
11	0	0	0	0	0	0	0	0	0	1	1	0	1	1	1
12	0	0	0	0	0	0	0	0	0	1	1	1	0	1	1
13	0	0	0	0	0	0	0	0	0	1	1	1	1	0	1
14	0	0	0	0	0	0	0	0	0	1	1	1	1	1	0

This matrix is used to determine the centrality which is the measure that will be explain in the next section.

4   Types of Centrality Measures

Centrality measure in social networks are frequently use to analyze the efficient communication, also the centrality characterized the behavior of communicating groups and identify the point that control its communications [7].

4.1   Degree Centrality

In social networks centrality was introduced by Bavelas in 1948, he proposed that there is a relationship between structural centrality and influence in group processes [8]. In 1954 Shaw introduce the degree as the value of centrality. Then the degree centrality is defined as the count of the number of nodes$_j$ $(i \neq j)$ that are adjacent to node$_i$. Degree centrality is a way to find the node that is strategical located to communicate or influence a group to propagate the information or not.

4.2   Eigenvector Centrality

Eigenvector centrality of a node$_i$ is determined by adjacency matrix applying Eq. 2. Where x_i' is defined by the sum of the i's centralities of I neighbors, and A_{ij} is an element of the adjacency matrix [9]. The interpretation of eigenvalues tells if a node is growing or shrinking according to the number of neighbors.

$$x_i' = \sum_j A_{ij} x_j$$

This metric is applying to capture the behavior of a network, hierarchy nodes, and detects interactions patterns.

4.3   Closeness Centrality

The closeness centrality of a node$_i$ is the inverse of the sum of the number of hops in the shortest paths from the node$_i$ to the rest of the nodes in the network [9]. This measure as not useful to discriminate or classified nodes in a network, at least in the

Table 3 Results of centrality measures of Barbell graph of Fig. 4

Nodes	Degree	Closeness	Betweenness
0	0.357142857	0.285714286	0
1	0.357142857	0.285714286	0
2	0.357142857	0.285714286	0
3	0.357142857	0.285714286	0
4	0.357142857	0.285714286	0
5	0.428571429	0.35	0.494505495
6	0.142857143	0.378378378	0.527472527
7	0.142857143	0.388888889	0.538461538
8	0.142857143	0.378378378	0.527472527
9	0.428571429	0.35	0.494505495
10	0.357142857	0.285714286	0
11	0.357142857	0.285714286	0
12	0.357142857	0.285714286	0
13	0.357142857	0.285714286	0
14	0.357142857	0.285714286	0

Barbell model many of the nodes that integrate the bells have the same value of closeness it can view in Table 3.

4.4 Betweenness Centrality

Betweenness centrality of a node the fraction of the shortest paths going through node k when considered over all pairs of nodes i and j [9]. Equation 3 define the betweenness of a node as follow.

$$Bc(k) = \sum_i \sum_j \frac{sp(i,k,j)}{sp(i,j)} \, i \neq j \neq k$$

where $sp(i,j)$ is the total number of shortest paths between nodes i and j, and $sp(i,k,j)$ is the number of shortest path that go through node k [1, 9]. This centrality metric helps to detect the role of a node in distributing information.

After all these definitions of centrality in Table 3 illustrate the results of applying the centralities of degree, closeness and betweenness of Barbell network, and Fig. 5 shows that betweenness metric is the indicate better the influence of a node to propagate a message.

Other way to analyze the structure of a network is to calculate different measures of centrality and graph it together to determine which centrality is best to solve the issue.

Complex networks approach is useful to analyze complex problems which cannot be solved by separation of its parts and then put together to get a solution.

Fig. 5 Relationship between centrality measures of degree, closeness and betweenness of barbell graph

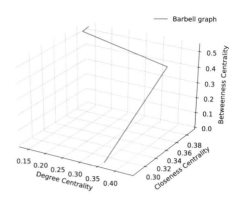

5 Agent-Based Modeling

An agent is an autonomous computational individual or object with preferences and actions, this is defined by a computational model. The implementation of agent-based models helps to understand the patterns behavior and emergence of a phenomenon [10]. In complex systems, the interactions of the elements are important features of emergence. The use of computational modeling as agent-based modeling (ABM) enables the simulations of complex systems.

The implementation of ABM allows us to model a set of agents inside of an environment and interact to each other, using a set of rules. Netlogo is a useful and easy language for the implementation of a simulation model. Other benefits of the use an ABM is that allows the programing of heterogeneous agents, in this way an history of interactions of each agent can be observed and found its strategy.

5.1 Proposed Agent-Based Model for Rumor Spreading

The spreading of a rumor depends of the structure of the network which represents the relations between the individuals that composed it. The modeling that we proposed use a set of agents that are in a classroom environment. Agent i is defined by a tuple $Ag_i = \langle d, c, b, s, f \rangle$ where d represents the degree centrality, c the closeness centrality, b the betweenness centrality, s is the status of cooperate that can be cero or one, and f this attribute represent the flow of information, this is set to one when the agent (student) got the message.

The environment where the agents are located is classroom that is integrate by three subgroups α, β, and λ. Figure 6. illustrate the ABM in which we will discover and analyze a set of agents, interconnected by a topology with the distribution of a Barbell graph. Our model calculates three types of centralities, in Fig. 7 the graph of the left shows the degree centrality, mean while the betweenness centrality is show on the right plot, and finally the closeness centrality at the bottom.

Fig. 6 Interface of the
agent-based model in Netlogo

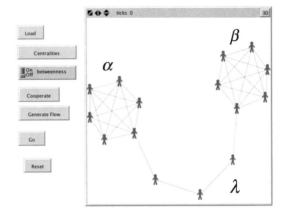

Fig. 7 Results of centralities
calculate by the model

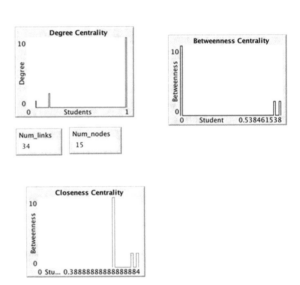

After the setup of the world that implements the creation of the topology by importing the adjacency matrix of Barbell graph, the next step is the initialization of the attributes of each agent (student). In Netlogo the properties of an agent are observed by inspecting each node in Fig. 8, turtle 9 is an agent with status cooperation set to cero.

The assignation of s (status of cooperate) is set in a random way, in our model agents with green color will cooperate to spread the rumor, and oranges agents not. Finally, the attribute f is set one by a random selection, this mean that only one agent will start the spreading of the rumor, in this case the color of the agent will set to yellow. Every ABM made in Netlogo has a process traditional called "go". To explain how this process works we must observe the sequence shows in Figs. 9, 10 and 11.

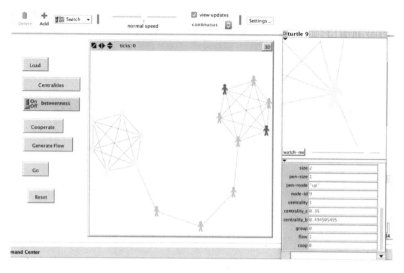

Fig. 8 No cooperation emerges in a pattern where only alpha group got the message

Fig. 9 Setup of the world and start of the rumor in agent number one with yellow color, orange color means no cooperate status, and green agents will cooperate to the spread of the rumor

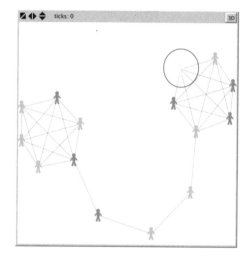

After one tick the distribution of the rumor depends of two attributes the first one is the status of cooperate and the second is his betweenness centrality. The rule will set if s == 1 the agent will spread the rumor only to his neighbors. Figure 10 shows the spreading of the rumor.

In Fig. 10 we can observe that one of the member of λ got the message and even all the members of β. The next step is determining if the rest of the agent will get the message or not. Figure 11 illustrate the last member of λ got the messages, but him will not propagate because his status of cooperate is cero. Finally, after three

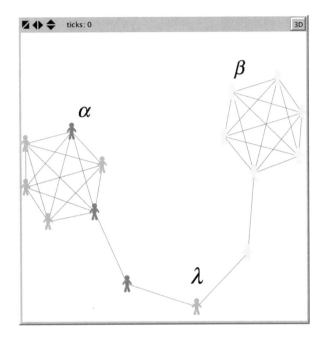

Fig. 10 The spreading of the rumor during the tick number two

Fig. 11 The last agent that got the message is inside of blue circle

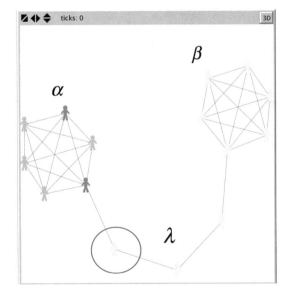

Fig. 12 Only the agent with the same betweenness centrality received the message

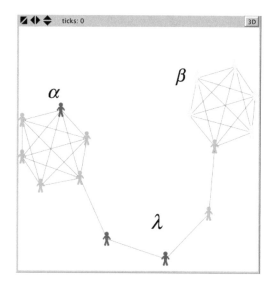

ticks the rumor was got for all the members of λ and β, understanding that no one of α group could get it.

In the ABM proposed we can activate the betweenness centrality button has another attribute to verify when the propagation of a rumor start. This scenery is shows in Fig. 12, where the rumor start in one of the member of β group and to spread the rumor only if him status of cooperate is set to one, and the betweenness of his neighbors is equal to himself.

6   Results

The configuration of the status of cooperate is the key for the propagation of a rumor when the betweenness centrality button is off. Figure 13 shows the status of each group and individuals (students) that received the message, the yellow color of the student mean that the rumor was gotten.

To spread the rumor to every agent that belong to the classroom is necessary observe Fig. 14, in this figure we identified that the members of λ group have status of cooperate set to one.

The λ group in a classroom have the control of the rumor distribution, when is necessary to broadcast a message in a group with Barbell topology.

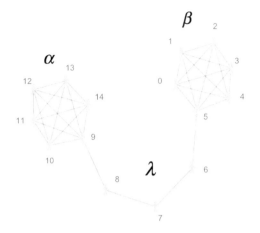

Fig. 13 Status of the network when everyone got the message

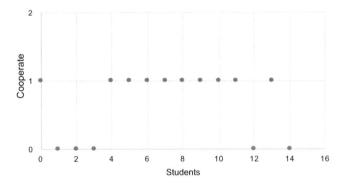

Fig. 14 Configuration of cooperation state of the agents (students)

7 Conclusions

The dynamics of rumor propagation are complex, as the interactions between the individuals that make up the classroom are observed. The simulation with intelligent agents allows the detection of emerging local and global behaviors, with only few study variables such as measures of centrality. The calculation of betweenness centrality in a network is relevant when distribution of information is analyzing. Finally, the centrality measure helps to the detection of leaderships in a group.

8 Future Work

As a future work is necessary to incorporate a traditional model for the spread of disease as SIR (Susceptible, Individuals, Recovered).

References

1. Estrada E (2011) The structure of complex networks: theory and applications. Oxford University Press Inc, New York
2. Barrat A, Barthélemy M, Pastor-Satorras R, Vespignani A (2004) The architecture of complex weighted networks. PNAS 101:3747–3752
3. Chung Fan, Lu L (2006) Complex graphs and networks (CBMS regional conference series in mathematics). American Mathematical Society, Boston
4. Watts DJ, Strogatz SH (1998) Collective dynamics of "small-world" networks. Nature 393:440–442. https://doi.org/10.1038/30918
5. Barabasi A-L, Pósfai M (2016) Network science graph theory. Cambridge University Press, Cambridge
6. Aldous D, Fill J (1999) Reversible Markov chains and random walks on graphs, vol 2002, pp 1–516
7. Freeman LC (1977) A set of measures of centrality based on betweenness. Sociometry 40:35–41
8. Freeman LC (1979) Centrality in social networks conceptual clarification. Soc Netw 1:215–239
9. Meghanathan N (2015) Use of eigenvector centrality to detect graph isomorphism. Comput Sci Inf Technol (CS IT) 5:01–09. https://doi.org/10.5121/csit.2015.51501
10. Wilensky U, Rand W (2015) An introduction to agent-based modeling

Multi-layered Network Modeled with MAS and Network Theory

Jose Parra, Carelia Gaxiola and Manuel Castañón-Puga

Abstract Complex networks have been widely used to model diverse real-world systems with great success. However, a limitation emerges when it comes to modeling real multilayered systems where nodes belong to different layers at the same time and have multiple interactions. Some of the research conducted under the multilayer approach focuses on the structure and topology of the network; to measure the resilience and robustness simulating the elimination of nodes in the network in this work, the theory of networks is used to model the problem of the European air transport network. Centrality measures will be applied to obtain information from the crucial entities Within a network and the topology formed layer by layer, multi-agent systems are applied to simulate the scenario of a negotiation when the risk of cancellation of flights is presented. By including multiagent systems, we intend to approximate a bit more to real systems since it allows to simulate multiple scenarios, provide them with mechanisms of negotiation and decision making.

Keywords Network theory · Multiagent systems · Complex systems

1 Introduction

The use of technology has drastically changed the way in which systems around the world interact and communicate, now they are highly interconnected, and there is high dynamism between them. In an increasingly complex world, numerous

J. Parra (✉) · C. Gaxiola · M. Castañón-Puga
Facultad de Ciencias Químicas e Ingeniería, Universidad Autónoma de Baja California,
Mexicali, Mexico
e-mail: jose.parra95@uabc.edu.mx

C. Gaxiola
e-mail: cgaxiola@uabc.edu.mx

M. Castañón-Puga
e-mail: puga@uabc.edu.mx

© Springer International Publishing AG, part of Springer Nature 2018
M. A. Sanchez et al. (eds.), *Computer Science and Engineering—Theory
and Applications*, Studies in Systems, Decision and Control 143,
https://doi.org/10.1007/978-3-319-74060-7_6

systems are composed of a large number of dynamic and highly interconnected units [1]. These complex systems do not have a unique definition accepted by the scientific community, however, in the various attempts to conceptualize it, some keywords are present, such as:

- Emerging Behavior.
- Decentralized.
- Auto Organization.
- Simple Elements.

Melanie Mitchell definition of complex systems is one of the most popular [2], she describes complex systems as extensive networks of elements lacking centralized control and obeying simple rules of operation, exhibiting complex collective behavior, sophisticated information, and adaptation by learning or evolution.

Through examples, we can have a precise idea of what a complex system is. The most common case at the research beginning is the study of ant colonies. Where hundreds or thousands of ants, each one with a simple behavior, with the task of finding food, react to a chemical signal from other ants in their colony, fight against other ants or protect their nest; as they cooperate to build complex structures or to carry heavy food and build their nest. Through observation, we can understand their behavior, adaptation to change, and how from their particular actions emerge a complex behavior when working in a group.

With the development of analysis tools, it is possible to study the interaction between dynamic units of technological, socioeconomic and biological systems, among others [3–6].

Some real network systems exhibit characteristics of heterogeneity. This heterogeneity may be related to the dynamics between their units, for example, diverse types of links in a social network [7–11], the variable electrical transmission capacity in a neural network [8, 12–14], and irregular data traffic on the web [15], to name just a few.

Other characteristics observed that have attracted the attention of the scientific community are robustness, tolerance to attacks and errors. Robustness in a network is the ability to prevent malfunction when a fraction of elements is damaged, and the tolerance to attacks and errors is the ability of the system to maintain connection properties after removal or disruption of a part of the nodes [16].

A fraction of the damaged nodes can lead to a cascade failure across the network, changing the connection flow and triggering a load redistribution across the network [17–24].

Sayama [25] has divided the study of complex systems into seven areas, as shown in Table 1, of which the following are discussed in this chapter:

- **Networks** (Graph Theory, Robustness/Vulnerability, Scale-Free Network, Social Network Analysis, Centrality, Small-World Networks)
- **Game Theory** (Rational decision making, negotiation, cooperation)
- **Collective Behaviour** (Agent-Based Modeling).

Table 1 Complex systems areas, with emergency and self organization as central concepts across all areas

Complex systems	
Emergency (over scale) and self-organization (over time)	
Networks	Graph theory, scaling, robustness/vulnerability, scale-free network, social network analysis, centrality, small-world networks, adaptive networks among others
Evolution and adaptation	Machine learning, artificial life, artificial neural networks, evolutionary computation, genetics algorithms/programming
Pattern formation	Spatial fractals, spatial ecology, self-replication, cellular automata, percolation among others
Systems theory	Entropy, computation theory, information theory, cybernetics, system dynamics, complexity measurement
Nonlinear dynamics	Attractors, chaos, bifurcation, time series analysis, ordinary differential equations
Game theory	Rational decision making, cooperation versus competition, evolutionary game theory, prisoner's dilemma (Pd)
Collective behavior	Social dynamics, collective intelligence, agent-based modeling, ant colony optimization among others

2   Related Work

Complex systems have been used to model real-world situations in which others approach seems to be insufficient.

2.1   *Complex Networks*

The natural framework for the mathematical treatment of complex networks is graph theory [16]. A graph is composed of nodes and links; it can be directed, in which case the order of the nodes is essential, or can be directed, another possible configuration is not directed with weights in its connections.

In Fig. 1 we observe three types of graphs with nodes and their links. Figure 1a represents a graph in which the interaction between the nodes is not directed, and the communication is given in two ways, back and forth. Figure 1b depicts a graph where its nodes are connected with directed links in one way, to have two nodes fully communicated it is necessary to have two links. Finally, in Fig. 1c is a graph with non-directed links, each one has a numerical value, commonly called weight.

With the graph model, we can observe characteristics of great importance in complex networks, such as:

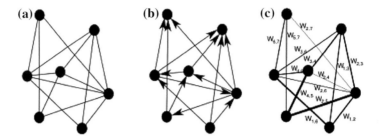

Fig. 1 Representation of graphs: **a** Graph with nodes and non-directed links. **b** Graph with nodes and directed links and **c** Graph with nodes and link weights

- **The degree of a node or connectivity**, given by

$$k_i = \sum_{j \in \mathcal{N}} a_{ij}. \tag{1}$$

  where k_i of a node i is the number of edges incident to with the node and is defined in terms of an adjacency matrix $\mathcal{A}$.

- **The average nearest neighbor degree** of a node i, given by

$$k_{nn,i} = \frac{1}{k_i} \sum_{j \in \mathcal{N}_i} k_j = \frac{1}{k_i} \sum_{j=1}^{N} a_{ij} k_j, \tag{2}$$

  where the sum is executed in the nodes belonging to $\mathcal{N}_i$ the set of the first neighbors of i.

- **Average shortest route** where d_{ij} is the length of the route from node i to node j, given by [26, 27]

$$L = \frac{1}{N(N-1)} \sum_{i,j \in N, i \neq j} d_{ij} \tag{3}$$

- **Clustering** can be quantified by defining the transitivity T of the graph as the relative number of transitive triples [28–30] given by

$$T = \frac{3 \times number\ of\ triangles\ in\ G}{Number\ of\ connected\ triples\ of\ vertices\ in\ G} \tag{4}$$

All these characteristics are essential to determine how connected the network is, to recognize which nodes have greater connectivity, which routes are the shortest or most critical, to identify subgroups in the network, among additional features of the system [16].

One advantage of complex network theory is the logical representation of the interconnection between an immense quantity of simple elements (as a complex system), this approach has been used to represent failures in cascade. Traditionally, when a multi-layered network is modeled, to simulate the communication, short paths, wrong paths, a random or directed attack one of the following methodologies, among others, is applied:

- *Small World Network.* This network model has a heterogeneous betweenness distribution, although its degree distribution is homogeneous. Further studies reveal that this small-world network is robust to random attack, but fragile to intentional attack when a failure in cascade scenario is presented [31].
- *Scale-free Network.* Mathematically, the power-law distribution means that statistical moments of the degree variable are not defined, hence the name of scale-free networks. Because of the ubiquity of scale-free networks in natural and human-made systems, the security of these networks, i.e., how failures or attacks affect the integrity and operation of the networks, has been of great interest since the discovery of the scale-free property [32].
- *Directed Network.* Most real-world networks, e.g., power grid, Internet, and transportation networks; can be modeled as directed networks, where the end to end path consist of a set of intermediate nodes from the source to the destination [33].

2.2 Multilayer Networks

Multi-layered networks constitute the natural environment for representing interconnected systems, where an entity may be present in more than one of these systems at the same time. Every one of these systems constitutes a layer of the network [34]. It is clear that nodes in some complex systems often have interactions of different types that take place over several networks that interact, that is, they constitute a multiplexed network. For example, in a social network [35], an individual has different interactions with other nodes such as family, peers, friends, and others, if each of these interactions is represented in distinct layers the model would be closer to reality [36].

Formal definition of a multilayer network

A multilayer network is a pair $\mathscr{L} = (\eta,\ E)$ where $\eta = \{N\alpha;\ \in \{1\ \ldots\ \mathscr{L}\}\}$ is a (Directed or Undirected, Weighted or Unweighted) graph $N\alpha = (X\alpha,\ I\alpha)$ (layers of $\mathscr{L}$) and

$$E = \{I_{\alpha\Phi} \subseteq X_\alpha \times X_\Phi; \alpha\Phi \in \{1\ldots\mathscr{L}\}, \alpha \neq \Phi\} \tag{5}$$

is the set of interconnections between nodes of different layers N_α and N_Φ with $\alpha \neq \Phi$. The elements of E are called crossed layers, and the elements of each I_α are

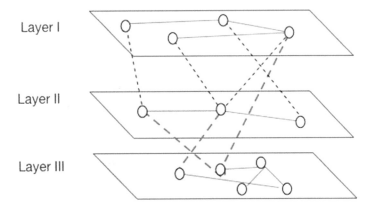

Fig. 2 Multilayer network with three layers

called intralayer connections of $\mathscr{L}$ in contrast with the elements of each $I_{\alpha\Phi}$ ($\alpha \neq \Phi$) that are called interlayer connections.

The concept of multilayer network extends to other mathematical objects [34] such as:

- Multiplex networks [37]
- Temporal networks [38]
- Interacting or interconnected networks [39]
- Multidimensional networks [40]
- Interdependent networks [41]
- Multilevel networks [42]
- Hypernetworks [43]

With the analysis of multilayer networks, the elements that are crucial in the structure of the system can be identified. The interactions can be classified, and the propensity of the nodes to form triangles can be discovered, as well as to recognize which node is the most used as a bridge for reaching other nodes, among other metrics that contribute to discern relevant information that could help to minimize contingencies in the network structure.

Figure 2 depicts a system with three layers represented by graphs, where a node has interactions with other nodes, some in its layer and some in another, while at one level it may be important for the operation, at a different level it can be a node with slight relevance.

There is some research on multilevel networks where the focus is the fault tolerance, the measures of centrality are the most used, and within them:

- **The degree of each node** is one of the most important, as a node have a higher quantity of connections, it becomes more relevant in the system. The degree of a node $i \in X$ of a multiplex network $\mathscr{L} = (\eta, E)$ is the vector [40, 44]

$$K_i = \left(K_i^{[1]} \dots K_i^{L} \right) \tag{6}$$

where K_i^{α} it is the node degree i in the layer α, this is $K_i^{[\alpha]} = \sum_j a_{ij}^{[\alpha]}$.

This formalization of the degree of the node is the natural extension of the degree of the node in the monolayer complex network.

- **Betweenness centrality** aims to calculate the appearing occurrence of a node in the geodesic path between two nodes.
- **Eigenvector** that measures the influence capacity of a node in the network, i.e., if is neighboring nodes with a high centrality degree, then its relevance is high on the network, and
- **Clustering coefficient** of the graph, given by Watts and Strogatz [26], can be extended from a monolayer network to the concept of a multilayer network in many ways. The measure of grouping gives us; as a result, the tendency of the nodes to form triangles, following the famous saying "the friend of my friend is my friend."

In a network $N(X, I)$ the clustering coefficient of a node i is given by

$$c_N = \frac{number\ of\ links\ between\ neighbors\ of\ i}{largest\ possible\ number\ of\ links\ between\ neighbors\ of\ i} \tag{7}$$

In [36], the European air transport network is modeled. The central objective is to reschedule flights to passengers, and above all, observe the network resilience to random failures. To simulate this problem a multilayer network was implemented, the airlines was modeled as layers, and the same nodes are present in all layers, flights were direct from the origin to destination, in their study they conclude that the multilevel structure increase considerably the resilience of the system against disturbances.

2.3 Multi-agent System

With the multiagent systems, the interaction between different entities represented by agents can be perceived; this is a significant feature in complex systems. Agent-based simulation has proven to be a tremendously useful technique for modeling complex systems, and especially social systems [45, 47–49]. Its objective is to find methods that allow us to build complex systems models, composed of autonomous agents that, while operating only with local knowledge and possess limited capacities, are capable of implementing the desired global behaviors [50].

Multi-agent systems are composed of several agents that interact with each other by exchanging messages, either to assist in a problem-solving assignment or to accomplish a series of individual or collective targets [45].

An agent is a system situated in a location with the ability of sense the environment in which is placed, and perform a series of actions to meet its objectives [46].

2.4 Multi-agent System Architectures

Different architectures have been proposed in order to develop agents with specific behaviors. According to the behavior, the agents can regularly be classified as deliberative, reactive and hybrid.

- *Deliberative* are those whose behavior can be predicted by the method of attributing beliefs, desires, and rational insight [51]. Among them, the intentional agents are the one of the most popular architectures agent BDI [52]. These agents are endowed with mental states such as beliefs, desires, and intentions, are based on the cognitive model of self-human so that agents use an internal representation of their surroundings, plus a symbolic model of the world around them.
- *Reactives* are elementary (and often memoryless) agents with a defined location in the environment. Reactive agents perform their actions as a consequence of the perception of stimuli coming either from other agents or from the environment; generally, the behavioral specification of this kind of agent is a set of condition-action rules, with the addition of a selection strategy for choosing an action to be carried out whenever more rules could be activated. In this case, the motivation for an action derives from a triggering event detected in the environment; these agents cannot be pro-active [52].
- *Hybrids* are the combination of the components of the reactive type, with components of deliberative type. Mostly used reactive part for interaction with the environment and the prompt reaction to events that may occur without investing time in reasoning, while the deliberative part is responsible for planning and decision making.

  The advantage of using this type of architecture is that it takes advantage of the other two types of architecture, and in fact, most of the phenomena have not adapted purely reactive or deliberative architectures. These architectures are typically organized in a hierarchical structure of layers, where the lower layers are primarily reactive and the higher layers are deliberative [46].

As mentioned above the BDI architecture is one of the most popular, it is based on the agent having a mental state as a basis for his reasoning. Figure 3 shows the interaction between the agent's reasoning and the environment.

- Beliefs are the agent's information about the world (environment), which does not necessarily have to correspond to reality. However, it is necessary for the agent to believe that the information is correct.

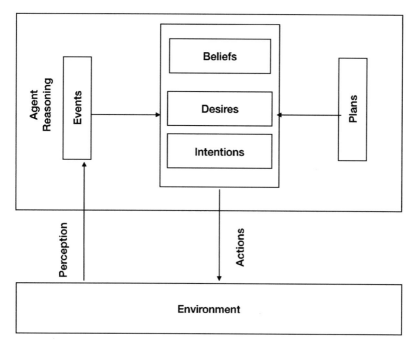

Fig. 3 Beliefs, desires and intentions architecture

- Desires are all possible task states that an agent wants to achieve. They represent the emotional state of the agent and could influence the agent's decisions.
- Intentions are the necessary plans to achieve a particular objective [53]. Plans are specifications of the steps to follow to achieve a goal.

The process of how the agent makes a decision is always something to analyze carefully. In [54] a review of 14 architectures that have achieved interest by the scientific community is presented, being the architecture belief-desire-intentions (BDI) one of the most popular [55], is recommended for building rationing systems for complex tasks in dynamic environments [56].

2.5 Negotiation

Negotiation is an advantageous technique used by agents when a goal or resource is necessary for more than one agent. There are four components required for a negotiation setting [57].

- Negotiation set.
- Protocol.

- Strategies, one for each agent.
- A rule to determine when an agreement is reached and when a deal is stuck.

One typical negotiation is the one with a series of rounds, with a proposal present in each round. There are three ways to carry out the negotiation between agents:

- One to one involves only two agents, and the negotiation is between them.
- Many to one, one agent negotiates with many other agents.
- Many to many, here negotiation is carried out simultaneously many agents with many other agents.

The one-to-one negotiation is the most commonly used, and the alternative offers one of most fundamental protocol one-to-one [58]. Figure 4 shows the basic model with two agents. Starting with the proposal of agent 1 and the acceptance or rejection of agent 2, if accepts means that agents have an agreement deal and the negotiation is finished, else agent 2 makes a proposal agent 1 can accept or reject, if accept then the negotiation ends, otherwise everyone starts over.

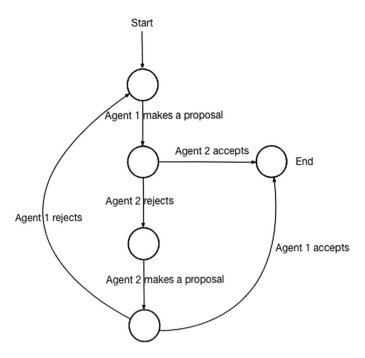

Fig. 4 The alternating offers model

3   Proposed Model

To obtain a model capable of analyzing the structure of the network, identify the most relevant entities, represent different objects with the ability to negotiate, make decisions and react to events, the use of an agent-based model in conjunction with network theory to model a multilayer system is proposed. Figure 5 shows the scheme of the proposal network theory and agent-based systems are applied to integrate complex systems into a single model.

In order to manage contingencies in multi-layer networks, this proposal is presented, with which we aim to represent entities capable of evaluating alternative solutions, negotiate resources and make decisions. The proposed model facilitates the analysis of the characteristics of the network structure, as well as the interaction and dynamics in its different layers.

Figure 6 shows agents that represent the nodes of the network and their links. Agents can be connected at the same time with agents inside and outside their layer, which is a characteristic of multilayer networks.

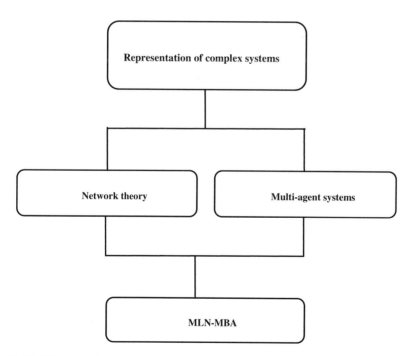

Fig. 5 Model proposed to represent complex systems, where network theory and multi-agent systems are used together

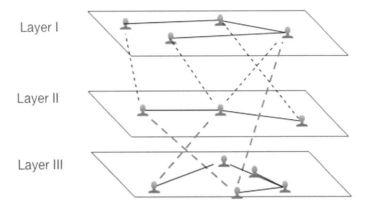

Layer I

Layer II

Layer III

Fig. 6 Multilayer network model

4 Case Study

The problem of the European transport network has been analyzed applying multiple approaches, in this work we propose the analysis of the network structure as a way to overcome the failures that arise when the functionality of the network nodes or their links is interrupted.

The data employed to build the network were used in [59]. First, the multilevel network was built; it has 37 levels $\mathcal{L}$, each representing an airline $\mathcal{A}$, the links i, j; ℓ represent the flights available from i to j. All airports are included in each layer, but only one airline. With the purpose of comparison, a monolayer network was built using the same data.

In the negotiation for the cancellation of flights, a set of agents $\mathcal{A} = \{\mathcal{Ag}_1, \ldots \mathcal{Ag}_n\}$ take part, with the tuple representing an agent is $\mathcal{Ag}_i = <\gamma, \tau>$, where γ represents the set of flight preferences $\gamma = \{f_1, \ldots f_{1n}\}$ and τ the finite set of available strategies to negotiate $\tau = \{s_1, \ldots s_{1n}\}$.

The negotiation after a flight cancellation will be one to one applying the protocol of alternative offers and will use the ultimate game at two rounds, where if they reach a conflict deal will result in a flight refund. The negotiation will be conducted in the following way: in the round 1, the passenger agent will say its offer, which the airline agent will accept or reject. If the airline agent accepts the offer, the negotiation is terminated. In the case of an offer rejection, the round 2 begins with a counteroffer from the airline, which the passenger can accept or reject. If the passenger rejects the counteroffer, Nash balance will not be reached, and this will trigger a conflict deal, so the passenger will seek to take what the airline agent proposes.

Table 2 describes the two-round negotiation process in an algorithm. Where they start by setting their preferences for the date to travel, and the strategies they will follow in the negotiation, the bid rounds restriction was established in two to allow both agents to propose offers. In this algorithm, no rule was established to indicate that it is better to accept any offer before reaching a conflict deal.

Table 2 Algorithm alternating offers protocol two rounds

function Ultimate Game

    set Ag_1, Ag_2 preferences

    set Ag_1, Ag_2 strategies

    Ag_1 makes an offer

    if Ag_2 accepts

        negotiation ends with an agreement

    else Ag_2 makes an offer

    if Ag_1, Accepts

        negotiation ends with an agreement

    else

        negotiation ends with a Conflict Deal θ.

end function

In both models, centrality measures will be applied as described below. Table 3 shows a comparison of the betweenness and centrality degree measures (the eigenvector measure is not included). The first column indicates the five high measures outcomes, and the next columns denote the name of the metric and the results of the monolayer network and the layer 17 and 20 of the multilayer network model.

The results both measurements present an apparent difference between the monolayer network and each layer of the multilayer network. Regarding the centrality degree, we can notice that in the monolayer network there are nodes with more than 100 connections, this is because the existence of the airlines is not being modeled.

If a highly interconnected node is eliminated the model could interpret that nobody else can make the flight, this is because the whole system could collapse for a single failure.

Table 3 Comparison of measures in a multilevel network and monolayer network

Position	Degree Layer 1	Degree Layer 17 of 37	Degree Layer 20 of 37	Betweenness Layer 1	Betweenness Layer 17 of 37	Betweenness Layer 20 of 37
1	112	40	41	2641	207	219
2	103	13	6	1339	148	214
3	100	3	5	1281	143	171
4	99	3	4	1130	84	144
5	95	2	4	859	76	115

On the other hand, the centrality degree of each layer of the multilayer network exhibits a lower connection due they represent only a fragment of the whole network, in the airlines' context, if a failure occurs in a layer, there is still the possibility of flying through a different layer (airline). Finally, from the measure of betweenness of the monolayer network showed in Table 3 we can infer that there are critical nodes in the network.

Figure 7 depicts a representation of the graph of airports and direct flights simulated as a monolayer network. Only airports and direct flights were taken into account.

To model the total loss of an airport the single-layer network would work well, but sometimes it is necessary to model the failure of one or more airlines and not the capacity of flights in general. In Fig. 8a, the modeling of the layer 17 is presented; it has 42 airports that carry out 53 direct flights, of which 20 flights can be operated by more than one airline. In Fig. 8b the layer 20 is depicted, it has 44 airports and 55 flights of which 25 can be done by more than one airline.

Some airports have a high passenger traffic, and this is the reason why they are considered an important node in the transport network. Figure 9a shows the measure of the centrality degree of the network of all airports with direct flights modeled as a single-layer network, in the chart some nodes (airports) have a high centrality degree, these are the airports with numerous connections, some of them having more than 100. In Fig. 9b depicts the results of the analysis of layer 17 of the multilevel network, the connections also represent direct flights between airports, as more direct flights are in the layer, the centrality of the node is higher. The results presented in Fig. 9b, c are lower than those in the chart of 9a because the analysis was performed on only one airline. The chart of 9c shows the results obtained by applying the measure to layer 20 when analyzing the chart(c) we

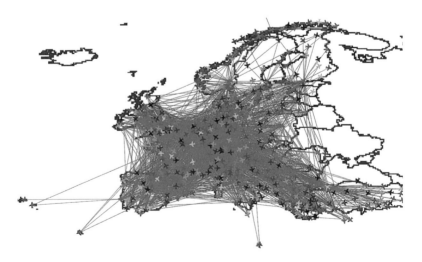

Fig. 7 Representation of European air transport networks as a monolayer

(a)

Layer 17 with 42 airports and 53 flights

(b)

Layer 20 with 44 airports and 55 flights

Fig. 8 A representation of European air transport networks multilayer network

observed that a node stands out remarkably from the others, with this we deduce that it is an essential node in the whole layer 20.

To recognize the busiest airports according to the connections, and which could be vulnerable to attacks due to their importance, betweenness is used. Figure 10a shows the measurement of betweenness of the monolayer network and Fig. 10b represents the results of layer 17 and Fig. 10c shows some nodes busiest of layer 20 of a multilayer network.

If a simple complex network was analyzed and the measurement of the eigenvector is calculated, the result will be the most influential nodes in the network. Figure 11a depicts the graph of the eigenvector measure a monolayer network, and Fig. 11b, c shows the graph of the eigenvector measure of a layer 17 and 20 from a multilayer network.

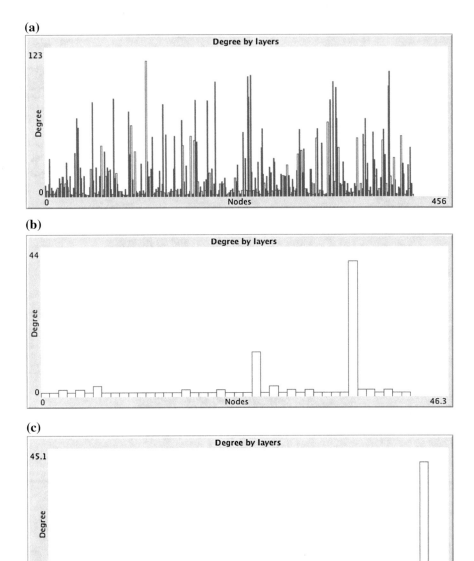

Fig. 9 Degree centrality of **a** monolayer network and **b** layer 17 and **c** layer 20

Table 4 shows the results of the negotiation in the layer 17, with 42 airports and 53 flights from the same airline. The results are from the average of 30 runs. In the column "Fly tomorrow" are those that according to their preferences could be scheduled for a flight the day after to avoid falling into a conflict deal, and the last column represents those who did not reach an agreement.

Fig. 10 Betweenness
centrality of network
a monolayer network **b** layer
17 and **c** layer 20 of
multilayer network

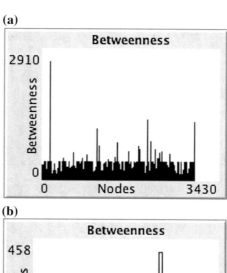

(a)

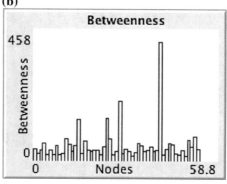

(b)

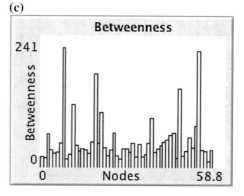

(c)

Table 5 shows the results of the negotiation in the layer 20, with 44 airports and 55 flights from the same airline. The results are from the average of 30 runs and five tests. In the column "Fly tomorrow" are those that according to their preferences could be scheduled for a flight the day after to avoid falling into a conflict deal, and the last column represents those who did not reach an agreement. The average obtained from the tests in the option to fly tomorrow is 27.6, and the average of those who chose a conflict is 17.4.

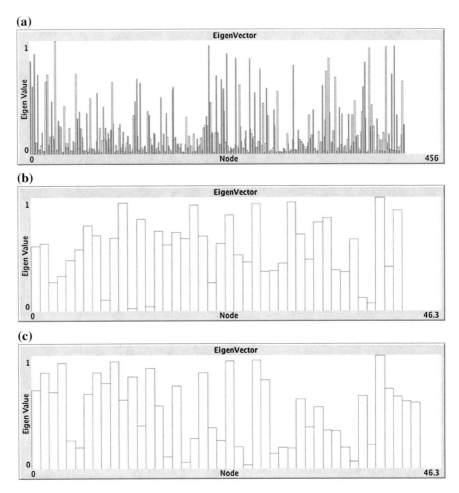

Fig. 11 Eigenvector measure of a network **a** monolayer network **b** layer 17 and **c** layer 20 of a multilayer network

Table 4 Negotiation alternative offers in the layer 17

Test	Fly today	Fly tomorrow	Conflict deal
1	10	26.46666667	16.53333333
2	10	26.5	16.5
3	10	27.3	15.7
4	10	26.46666667	16.53333333
5	10	27.06666667	15.93333333
Average	10	26.76	16.24

Table 6 shows the results of the negotiation of the monolayer network, which contains 417 airports and 2953 flights. The average of 30 executions made in 5 tests is shown. The limit in which the airline gives up in round 1 is shown in the first

Table 5 Negotiation alternative offers in the layer 20

Test	Fly today	Fly tomorrow Layer 20	Conflict deal
1	10	27.5	17.5
2	10	28.16666667	16.83333333
3	10	27.56666667	17.43333333
4	10	27.76666667	17.23333333
5	10	27	18
Average	10	27.6	17.4

Table 6 Negotiation alternative offers in a monolayer model

Test	Fly today	Fly tomorrow	Conflict Deal
1	600	1481.066667	871.9333333
2	600	1480.033333	872.9666667
3	600	1473.233333	879.7666667
4	600	1479.066667	873.9333333
5	600	1472.533333	880.4666667
Average	600	1477.186667	875.8133333

Table 7 Average results of the applied negotiation in monolayer and multilayer network

Network	Fly tomorrow	Conflict deal
Monolayer	1477.186667	875.8133333
Multilayer	26.76	16.24

column, in the second column are the times when the passengers do not want to fall in a conflict deal and accept the airline's offer, and in the last column the times that an agreement was not reached.

Table 7 shows the averages of the tests performed on the monolayer network and layers 17 and 20 of the multilayer network. In the column "Fly Tomorrow" show most of the passengers accepting the airline's treatment to avoid a conflict. In column "Conflict Deal" the average of the passengers who accepted the conflict deal.

5 Results

When airports and flights are simulated in the models presented in this work, we can conclude that a monolayer network considers a route as unique, i.e., in a case of any contingency the entire route is disabled, nobody can flight from origin to destination. Instead, the multilevel network manages the routes for each airline individually, and contingency affects the airline and not to the whole route.

When node degree measure is applied, the node with the highest connection is recognized. In our case of study, from the monolayer network, this identify the most connected airport; in the multilevel network, each level has an airport highly connected, this means that a node can be relevant in one layer but could not be significant in other layers.

To represent problems more attached to reality is necessary to simulate flights cancellations on a daily basis, in the simulation we observed that most passengers, agents that simulate negotiations with the airline, prefer not to fall into conflict and accept the airline's offer.

6   Conclusion and Future Work

With the simulations, it can be observed that modeling the problem of the European air transport network in a multilayer network shows us relevant information, that we could not easily be observed if the problem is modeled as a monolayer network. In case of a contingency, it tells us which airports are the busiest and which have more connections.

The use of agents allows us to model the problems with more characteristics that bring us closer to simulating real problems. The reactive agent that was implemented negotiates only if a flight cancellation is presented. However, it can be implemented to react to more complicated situations, such as flight delays, flight reschedule, choice of connections, among others.

As future work, it is necessary to include in the simulation agents to manage the different types of passengers, the flight costs, among other variables that affect the negotiations with the passengers of a canceled flight. Also, it would be necessary to carry out a network robustness test, as well as to calculate the tolerance to disconnections.

Acknowledgements We thank the MyDCI program of the Division of Graduate Studies and Research, Autonomous University of Baja California, Mexico, the financial support provided by our sponsor CONACYT contract grant number: 263888.

References

1. Bar-Yam Y (2013) Dynamics of complex systems. In: Dynamics of complex systems
2. Toroczkai Z (2010) Complexity: a guided tour complexity. Melanie Mitchell. Oxford University Press, New York, 2009 (349 pp). ISBN: 978-0-19-512441-5. Phys Today 63:47–48. https://doi.org/10.1063/1.3326990
3. Jeong H, Tombor B, Albert R et al (2000) The large-scale organization of metabolic networks. Nature 407:651–654

4. Jeong H, Mason SP, Barabasi A-L, Oltvai ZN (2001) Lethality and centrality in protein networks. Nature 411:41–42
5. Sole RV, Montoya M (2001) Complexity and fragility in ecological networks. Proc R Soc B Biol Sci 268:2039–2045. https://doi.org/10.1098/rspb.2001.1767
6. Camacho J, Guimerà R, Nunes Amaral LA (2002) Robust patterns in food web structure. Phys Rev Lett 88:228102. https://doi.org/10.1103/PhysRevLett.88.228102
7. Marchiori M, Latora V (2000) Harmony in the small-world. Phys A Stat Mech its Appl 285:539–546
8. Latora V, Marchiori M (2001) Efficient behavior of small-world networks. Phys Rev Lett 87:198701. https://doi.org/10.1103/PhysRevLett.87.198701
9. Newman MEJ (2001) Scientific collaboration networks. I. Network construction and fundamental results. Phys Rev E 64:16131. https://doi.org/10.1103/PhysRevE.64.016131
10. Granovetter MS (1973) The strength of weak ties. Am J Sociol 78:1360–1380. https://doi.org/10.1086/225469
11. Newman MEJ (2001) The structure of scientific collaboration networks. Proc Natl Acad Sci 98:404–409. https://doi.org/10.1073/pnas.021544898
12. Latora V, Marchiori M (2003) Economic small-world behavior in weighted networks. Eur Phys J B 32:249–263. https://doi.org/10.1140/epjb/e2003-00095-5
13. Sporns O, Tononi G, Edelman G (2000) Connectivity and complexity: the relationship between neuroanatomy and brain dynamics. Neural Netw 13:909–922. https://doi.org/10.1016/S0893-6080(00)00053-8
14. Sporns O (2002) Network analysis, complexity, and brain function. Complexity 8:56–60. https://doi.org/10.1002/cplx.10047
15. Romualdo Pastor-Satorras AV (2004) Evolution and structure of the internet. Cambridge University Press, Cambridge. https://doi.org/10.1017/cbo9781107415324.004
16. Boccaletti S, Latora V, Moreno Y et al (2006) Complex networks: structure and dynamics. Phys Rep 424:175–308. https://doi.org/10.1016/j.physrep.2005.10.009
17. Mantegna RN, Stanley HE, Chriss NA (2000) An Introduction to econophysics: correlations and complexity in finance. Phys Today 53:70. https://doi.org/10.1063/1.1341926
18. Watts DJ (2002) A simple model of global cascades on random networks. Proc Natl Acad Sci 99:5766–5771. https://doi.org/10.1073/pnas.082090499
19. Moreno Y, Vazquez A (2002) The Bak-Sneppen model on scale-free networks. Europhys Lett 57:765–771. https://doi.org/10.1209/epl/i2002-00529-8
20. Bak P, Sneppen K (1993) Punctuated equilibrium and criticality in a simple model of evolution. Phys Rev Lett 71:4083–4086. https://doi.org/10.1103/PhysRevLett.71.4083
21. Jensen HJ, Magnasco MO (1999) Self-organized criticality: emergent complex behavior in physical and biological systems
22. de Arcangelis L, Herrmann H (2002) Self-organized criticality on small world networks. Phys A Stat Mech Appl 308:545–549. https://doi.org/10.1016/S0378-4371(02)00549-6
23. Goh K-I, Lee D-S, Kahng B, Kim D (2003) Sandpile on scale-free networks. Phys Rev Lett 91:148701. https://doi.org/10.1103/PhysRevLett.91.148701
24. Caruso F, Latora V, Rapisarda A, Tadić B (2005) The Olami-Feder-Christensen Model on a small-world topology. In: Complexity, metastability and nonextensivity, pp 355–360
25. Sayama H (2015) Introduction to the modeling and analysis of complex systems
26. Watts DJ, Strogatz SH (1998) Collective dynamics of "small-world" networks. Nature 393:440–442. https://doi.org/10.1038/30918
27. Watts DJ (Princeton S in C (2000) Small worlds: the dynamics of networks between order and randomness. Princeton University Press, 1999, 262 pp. ISBN: 0-691-00541-9. Bull Math Biol 62:794–796. https://doi.org/10.1006/bulm.1999.0175
28. Newman MEJ The structure and function of complex networks
29. Barrat A, Weigt M (1999) On the properties of small-world network models
30. Alon N, Yuster R, Zwick U (1997) Finding and counting given length cycles. Algorithmica 17:209–223. https://doi.org/10.1007/BF02523189

31. Xia Y, Fan J, Hill D (2010) Cascading failure in Watts-Strogatz small-world networks. Phys A Stat Mech Appl 389:1281–1285. https://doi.org/10.1016/j.physa.2009.11.037

32. Wang J, Rong L, Zhang L, Zhang Z (2008) Attack vulnerability of scale-free networks due to cascading failures. Phys A Stat Mech Appl 387:6671–6678. https://doi.org/10.1016/j.physa.2008.08.037

33. Fang X, Yang Q, Yan W (2014) Modeling and analysis of cascading failure in directed complex networks. Saf Sci 65:1–9. https://doi.org/10.1016/j.ssci.2013.12.015

34. Boccaletti S, Bianconi G, Criado R et al (2014) The structure and dynamics of multilayer networks. Phys Rep 544:1–122. https://doi.org/10.1016/j.physrep.2014.07.001

35. Wasserman S, Faust K (1994) Social network analysis: methods and applications

36. Cardillo A, Zanin M, Gómez-Gardeñes J et al (2013) Modeling the multi-layer nature of the European air transport network: resilience and passengers re-scheduling under random failures. Eur Phys J Spec Top 215:23–33. https://doi.org/10.1140/epjst/e2013-01712-8

37. Solá L, Romance M, Criado R et al (2013) Eigenvector centrality of nodes in multiplex networks. Chaos An Interdiscip J Nonlinear Sci 23:33131. https://doi.org/10.1063/1.4818544

38. Holme P, Saramäki J (2013) Temporal networks. In: Physics reports

39. Donges JF, Schultz HCH, Marwan N et al (2011) Investigating the topology of interacting networks. Eur Phys J B 84:635–651. https://doi.org/10.1140/epjb/e2011-10795-8

40. Berlingerio M, Coscia M, Giannotti F et al (2011) Foundations of multidimensional network analysis. In: 2011 International conference on advances in social networks analysis and mining, IEEE, pp 485–489

41. Buldyrev SV, Parshani R, Paul G et al (2010) Catastrophic cascade of failures in interdependent networks. Nature 464:1025–1028. https://doi.org/10.1038/nature08932

42. Criado R, Flores J, García del Amo A et al (2012) A mathematical model for networks with structures in the mesoscale. Int J Comput Math 89:291–309. https://doi.org/10.1080/00207160.2011.577212

43. Criado R, Romance M, Vela-Pérez M (2010) Hyperstructures, a new approach to complex systems. Int J Bifurc Chaos 20:877–883. https://doi.org/10.1142/S0218127410026162

44. Battiston F, Nicosia V, Latora V (2014) Structural measures for multiplex networks. Phys Rev E Stat Nonlin Soft Matter Phys 89. https://doi.org/10.1103/physreve.89.032804

45. Gilbert N, Troitzsch K (2005) Simulation for the social scientist

46. Nolfi S (2002) Power and limits of reactive agents. Neurocomputing 42:119–145. https://doi.org/10.1016/S0925-2312(01)00598-7

47. Conte R, Hegselmann R, Terna P (1997) Simulating social phenomena. Springer, Berlin, Heidelberg

48. Gilbert N, Terna P (2000) How to build and use agent-based models in social science. Mind Soc 1:57–72. https://doi.org/10.1007/bf02512229

49. de Marchi S, Page SE (2014) Agent-based models. Annu Rev Polit Sci 17:1–20. https://doi.org/10.1146/annurev-polisci-080812-191558

50. Vidal J (2007) Fundamentals of multiagent systems with NetLogo examples. AI Mag 1:151. https://doi.org/10.4018/jats.2009071006

51. Dennett DC (1989) The intentional stance. Philos Books 30:169–172. https://doi.org/10.1111/j.1468-0149.1989.tb02170.x

52. Rao AS, Georgeff MP, et al (1995) BDI agents: from theory to practice. In: ICMAS, pp 312–319

53. Cohen PR, Levesque HJ (1990) Intention is choice with commitment. Artif Intell 42:213–261. https://doi.org/10.1016/0004-3702(90)90055-5

54. Balke T, Gilbert N (2014) How do agents make decisions? A survey. JASSS 17:1. https://doi.org/10.18564/jasss.2687

55. Georgeff M, Pell B, Pollack M et al (1998) The belief-desire-intention model of agency. In: 5th International workshop on agent theories, architectures, and languages, ATAL'98, pp 1–10. https://doi.org/10.1007/3-540-49057-4_1

56. Bordini RH, Hübner JF, Wooldridge M (2007) Programming multi-agent systems in AgentSpeak using Jason

57. Wooldridge M (2009) An introduction to MultiAgent systems, 2nd edn. Wiley
58. Osborne MJ, Rubinstein A (1990) Bargaining and markets
59. Cardillo A, Gómez-Gardeñes J, Zanin M et al (2013) Emergence of network features from multiplexity. Sci Rep 3:1344

A Fuzzy Inference System and Data Mining Toolkit for Agent-Based Simulation in NetLogo

**Josue-Miguel Flores-Parra, Manuel Castañón-Puga,
Carelia Gaxiola-Pacheco, Luis-Enrique Palafox-Maestre,
Ricardo Rosales and Alfredo Tirado-Ramos**

Abstract In machine learning, hybrid systems are methods that combine different computational techniques in modeling. NetLogo is a favorite tool used by scientists with limited ability as programmers who aim to leverage computer modeling via agent-oriented approaches. This paper introduces a novel modeling framework, JT2FIS NetLogo, a toolkit for integrating interval Type-2 fuzzy inference systems in agent-based models and simulations. An extension to NetLogo, it includes a set of tools oriented to data mining, configuration, and implementation of fuzzy inference systems that modeler used within an agent-based simulation. We discuss the advantages and disadvantages of integrating intelligent systems in agent-based simulations by leveraging the toolkit, and present potential areas of opportunity.

J.-M. Flores-Parra (✉) · M. Castañón-Puga · C. Gaxiola-Pacheco · L.-E. Palafox-Maestre
Facultad de Ciencias Químicas e Ingeniería, Universidad Autónoma de Baja California,
Tijuana, Baja California, México
e-mail: mflores31@uabc.edu.mx

M. Castañón-Puga
e-mail: puga@uabc.edu.mx

C. Gaxiola-Pacheco
e-mail: cgaxiola@uabc.edu.mx

L.-E. Palafox-Maestre
e-mail: lepalafox@uabc.edu.mx

R. Rosales
Facultad de Contaduría y Administración, Universidad Autónoma de Baja California,
Tijuana, Baja California, México
e-mail: ricardorosales@uabc.edu.mx

A. Tirado-Ramos
University of Texas Health Science Center at San Antonio, San Antonio, TX, USA
e-mail: tiradoramos@uthscsa.edu

© Springer International Publishing AG, part of Springer Nature 2018　　　　127
M. A. Sanchez et al. (eds.), *Computer Science and Engineering—Theory
and Applications*, Studies in Systems, Decision and Control 143,
https://doi.org/10.1007/978-3-319-74060-7_7

1 Introduction

Agent-based simulation is an increasingly common approach for modeling complex phenomena using computational science to approach complexity [1, 2]. However, complex systems require realistic models to approach real problems. Consequently, novel tools for the modeling and simulation of complex problems, which may help modeling realistic models, are needed. Realistic models are not trivial to build, though, since new computational techniques for complex-systems are required to rethink the problem. One feasible approach is, e.g. to extend the functionality of currently available software tools, adding the new required functionality.

That is, many currently available software tools may offer mechanisms to add new features, providing an Application Programming Interface (API) to developers so they may build such extensions to the base application framework.

NetLogo is an accepted agent-oriented tool used mainly by social scientists with limited ability to program a computer [3]. Although the built-in models in NetLogo may be simple representations of complex problems, the available analysis is instrumental in demonstrating newly hidden behaviors. As a result, the tool offers a simulation engine that has been widely adopted for the computational modeling of social issues [4]. This tool offers attractive features of functionality, practicality, and user-friendliness, and has grown organically as user features are added by the community. Very importantly, it is available as freeware.

Prediction capabilities are important when dealing with simulation of complex systems [5]. Machine Learning (ML) is a computational method used to develop models and algorithms that allow for prediction. There are different ML techniques could be used for getting a model from real data, for example, association rule learning, artificial neural networks, clustering, and so forth. Furthermore, Hybrid Intelligent Systems (HIS) are a recent addition to the tool arsenal for modeling complex systems. They are methods that combine different computational intelligence techniques in modeling, including Artificial Intelligence (AI) and Machine Learning technologies. Neuro-fuzzy systems and evolutionary neural networks are prime examples of HIS.

In this paper, we present JT2FIS NetLogo; this is a Toolkit for Fuzzy Inference Systems (FIS) integration in agent-based models and simulations. FIS is a Hybrid Intelligent Systems that implements a Type-1 or Type-2 fuzzy system as a Machine Learning mechanism for prediction. This is a versatile tool with which a researcher could, for instance, use projection as a Decision-Making system in Agent-Based Modeling and Simulation.

The simplicity of use offered by NetLogo was part of the rationale for us to choose it as an appropriate platform in which to include fuzzy systems, and in particular the more innovative Type-2 Fuzzy Inference Systems (T2FIS) generation. Aside from providing this extended toolkit, we intend to show the usefulness of Type-2 Fuzzy Inference Systems in representing realistic social settings, as well as contribute to the improvement of available computational models.

Furthermore, in this paper we also discuss the advantages and disadvantages of integrating intelligent systems in agent-based simulations, and present some potential areas of opportunity and relevant applications.

1.1 Fuzzy Logic as a Methodology

The main contribution of Fuzzy Logic(FL) to this particular problem is a methodology for computing by using words [6]. One of our aims, in the narrow sense, is to show that fuzzy logic has well-developed formal foundations as a logic of imprecise (i.e. vague) propositions, and that most events that may be named "fuzzy inference" can be naturally understood as logical deductions [7].

Fuzzy Logic remains quite an active line of research [8]. According to Wan [9], its principal objective is to use fuzzy set theory for developing methods, concepts for representing and dealing knowledge expressed by natural language statements. A fuzzy system inference (FIS) is able, therefore, to understand the sum of a system of inference. A FIS is a rule-based classification method that uses fuzzy logic to map an input space into an output space.

FIS is based on three components: a rule base containing a set of fuzzy if-then rules, a database that defines the membership functions used in the rules, and a reasoning mechanism that performs the inference process consisting of applying the rules to achieve a specific result [10]. Mamdani FIS and Takagi-Sugeno FIS are types of inferences systems. Both have *IF-THEN* rules and the same antecedent structures. However, there are differences between them. Firstly, while the structure of the consequents for a Takagi-Sugeno FIS's rule is a function, in a Mamdani FIS's rule is a fuzzy set [11]. Secondly, defuzzification method is necessary for Mamdani FIS to obtain the crisp output because the output of a Mamdani FIS is a fuzzy set, whereas the output of a Takagi-Sugeno is a crisp value. Thirdly, the most significant difference in Takagi-Sugeno FIS is that it is computationally effective but loses linguistical interpretability for humans, whereas Mamdani FIS is intuitive and suitable to human interpretation [12].

Another useful concept is data mining that combines techniques from visualization, database, statistics, machine learning and recognition pattern methods with the objective of extracting and explain large volume of data. The primary goals of data mining are discovery, forecast and prediction of possible actions, with some factor of error per prediction [13]. Also, Data Mining helps to take decisions from identifying patterns, relationships and dependencies for generating predictive models [14].

Fuzzy C-Means (FCM) and Subtractive Clustering are also quite popular methodologies nowadays. Both of them can be used to extract patterns of data and create the initial configuration of an FIS. The Fuzzy C-Means algorithm (FCM) [15, 16] is basically the joining of c-means clustering algorithm with fuzzy data. This joining takes into account the data's uncertainty, helping prevent incorrect results and make correct crisp partitions [15].

A related methodology is that of Subtractive Clustering. It defines a cluster center based on the density of surrounding data [11], calculating the best choice of center based on mathematical approximations. The center data is determined by the distance of this data point from all other data points [17]; cluster center can represent a fuzzy rule of the system, and each found group represents the antecedent of this rule.

Finally, Agent-based modeling (ABM) is a simulation modeling technique which has greatly developed in last few years, e.g. in applications ranging from business to social problems [18]. ABM are commonly used to approach a complex system utilizing agents as main elements [19]. In ABM, problems are modeled through autonomous agents, capable of decision making. Each agent can evaluate its situation and make decisions by a set of rules. Further, agents have behaviors appropriate for the system they represent. These behaviors may also be based on interactions between agents, a main characteristic of agent-based modeling [18], where the agents interact with each other to simulate complex environments and predict emergent behavior [20].

1.2   Related Work

A popular methodology nowadays, we next mention a few representative examples of FIS in agent-based modeling. In [21], fuzzy sets are used in agent-based simulations to represent emotions and fuzzy rules in order to describe how some events are triggered by employing emotions, and how these emotions produce different behaviors. In [22, 23] the authors propose the use of fuzzy logic to formalize different types of personality traits for human behavior simulation. In [24, 25] the authors use fuzzy sets in the context of trust and reputation. In [26] the authors formalize various measures of success as fuzzy sets in a spatially iterated Prisoner's Dilemma, and explore the consequences. In [27] propose the use of fuzzy logic in Social Simulation to formalize concepts such as conflict, violence, and crime. In [28, 29] the authors incorporate the interpolation method in the decision making of their agents. In [30, 31] the authors explore discussion dynamics of competing products in different markets using agent-based models where various linguistic terms are formalized as fuzzy sets.

2   The JT2FIS NETLOGO Tool-Kit

The main idea behind the development of our JT2FIS NETLOGO toolkit framework is to help researchers implement Type-1 and Type-2 Fuzzy Inference Systems in NetLogo models, in particular where agents take decisions or express preferences through diffuse logic.

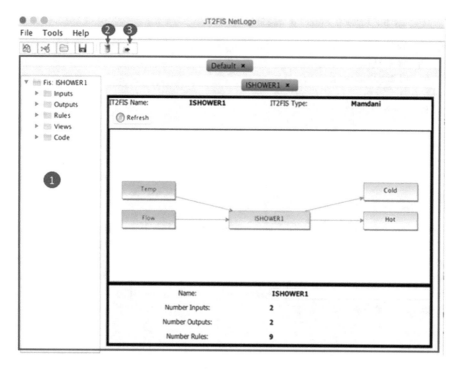

Fig. 1 Main elements for JT2FIS NETLOGO Tool-kit

The JT2FIS NETLOGO toolkit's architecture can be divide into three main elements (Fig. 1), as follows:

- Develop Mamdani and Takagi-Sugeno Fuzzy Logic System.
- Clustering.
- Export NetLogo.

2.1 Develop Mamdani and Takagi-Sugeno Fuzzy Logic System

The primary objective of this feature is to provide researchers with the integration of diffuse inference systems such as Mamdani and Takagi-Sugeno to their models. Also, it facilitates the visualization and configuration of the FIS to be able to include them in the simulations based on agents. The central core of this tool is basing on the JT2FIS library proposed in [32].

JT2FIS NETLOGO toolkit has five main modules for developing FIS:

1. Inputs.
2. Outputs.
3. Members Functions.
4. Rules.
5. Data Evaluation.

To explain how to create these fuzzy inference systems, we will use the ISHOWER example of Matlab.

2.1.1 Inputs

Using this module, we can add and manage the inputs (input linguistic variables values) of our diffuse inference system. The primary attributes of the inputs are the name of the variable and the limit of the values that can take (lower and upper limit). Figure 2 we show the Temperature entry for the ISHOWER example and its configuration (Name, upper and lower range, etc.) in addition to the membership functions that form it.

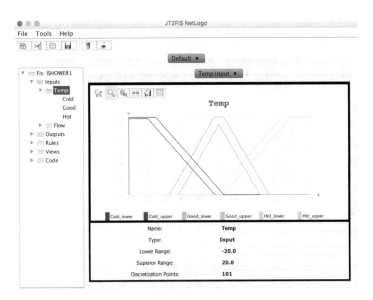

Fig. 2 Configuration of "Temp" input in "ISHOWER" example

2.1.2 Outputs

The output module allows us to add and modify the characteristics of the outputs (output linguistic variables values) of our FIS. As in the inputs, the principal attributes of the outputs are the name of the variable and its limit (lower and upper limit). Figure 3 we show the output "Cold" of the example ISHOWER and its configuration (Name, upper and lower ranger, etc.) in addition to the membership functions that compose it.

2.1.3 Members Functions

Membership functions are the linguistic values that an input or output linguistic variable can take. The membership functions used in our tool can see in [32]. In this module, you can add, delete and modify the attributes of membership functions as their name, parameters that compose it. Figure 4 shows the linguistic variable "Good" of the entry "Temp" of the example ISHOWER FIS this is of the triangular type with uncertainty in all the sides that compose it.

2.1.4 Rules

The rules are the basis of the knowledge of the diffuse Inference System. In this module, we can add, edit and delete these rules. This module shows us the inputs

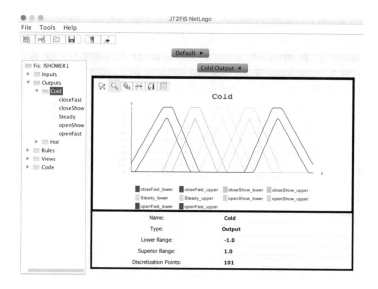

Fig. 3 Configuration of "Cold" output in "ISHOWER" example

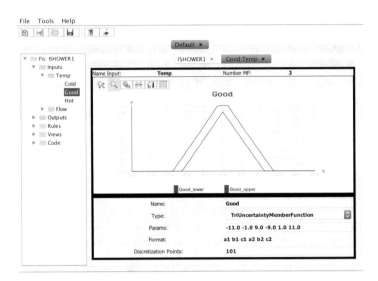

Fig. 4 Configuration of "Good" member function for "Temp" input in "ISHOWER" example

and outputs that make up the FIS; it is here where we must select the antecedents and consequents that make up each rule. The antecedents may be connected by the logical operators AND and OR. Figure 5 we can observe in more detail the formation of these rules.

2.1.5 Data Evaluation

Once the FIS configuration is complete, we can evaluate the data. When we talk about evaluating the data, we refer to giving numerical input values to its inputs. FIS through your system of knowledge (Rules) will provide us with the corresponding results for each of its outputs.

Figure 6 shows the FIS evaluation module. This module has different defuzzification methods; these methods are responsible for transforming the Fuzzy output to a numerical value. Also, this module offers other features such as configuration for methods implication, aggregation, and, or for inference.

We have two different options to select points to be evaluated. You can evaluate manually add points as shown in Fig. 7 or you can import a CSV file to evaluate a set of points automatically. The JT2FIS NETLOGO Tool-kit can export the results in CSV format.

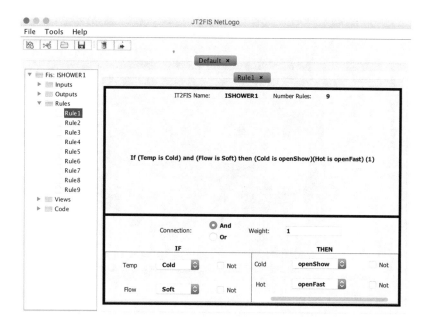

Fig. 5 Configuration of Rule 1 in "ISHOWER" example

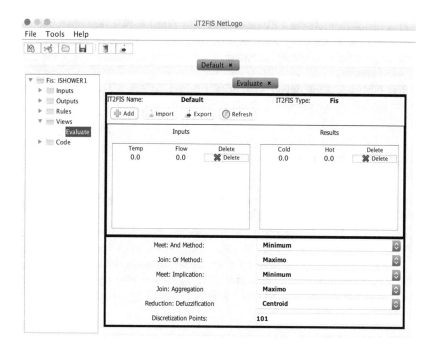

Fig. 6 Evaluate "ISHOWER" example

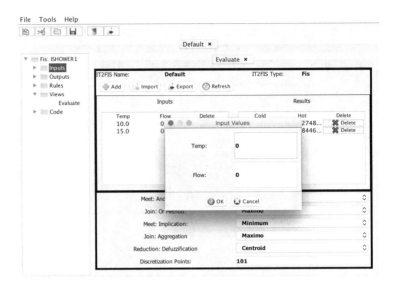

Fig. 7 Add point evaluate "ISHOWER" example

2.2 Clustering

The idea is to generate an FIS from CSV file through different clustering techniques. Figure 8 shows the graphic user interface of this panel. In this way, the user can set-up generation member functions and clustering method to apply the desired generation process.

Fuzzy c-Means is the default clustering method. In Table 1 list Type-2 member function available in JT2FIS NETLOGO Tool-kit.

2.3 Export NetLogo

For export the fuzzy inference systems generated, it is necessary to have the JT2FISNetLogo extension. JT2FISNetlogo is a NetLogo extension based on JT2FIS library. JT2FISNetlogo extension allows Fuzzy Inference Systems programming that can be accessible to non-specialists. The extension was built and tested on Version 5.3. The extension is available for educational purposes in [33].

The code generated for implementation of an FIS in NetLogo is saved in a file with extension .nls to be able to use it within any NetLogo model. Figure 9 shows an example of the generated code.

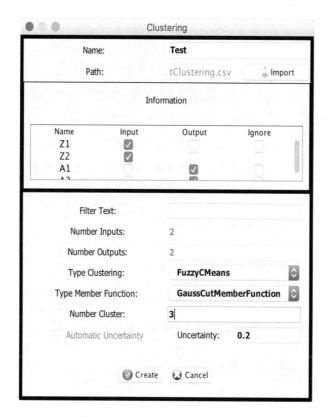

Fig. 8 Clustering panel user interface

Table 1 JT2FISCLUSTERING member functions

Type clustering	Type-2 member functions	Description
Fuzzy c-means	GaussCutMemberFunction	Params = [inputs outputs uncertainty]
Fuzzy c-means	GaussUncertaintyMeanMemberFunction	Params = [inputs outputs] Params = [inputs outputs uncertainty]
Fuzzy c-means	GaussUncertaintyStandardDesviationMemberFunction	Params = [inputs outputs] Params = [inputs outputs uncertainty]

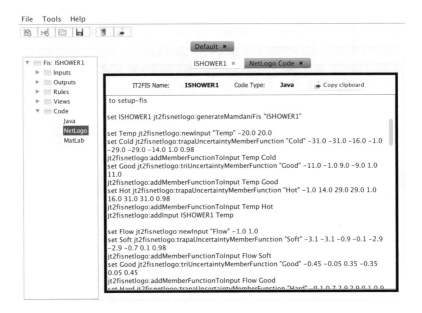

Fig. 9 NetLogo code generated in "ISHOWER" example

2.3.1 Executing FIS in NetLogo

To use JT2FISNetLogo extension, put the JT2FISNetLogo folder in the NetLogo extensions folder, or in the same directory as the model that will use the extension. At the top of the Code tab write: **extensions [jt2fisnetlogo]**.

The next step is to import the file with .nls extension. To import it is necessary to write down the previous line: **__includes["pathFile.nls"]**.

Finally, we evaluate the FIS with the next fraction of code **set outputList jt2fisnetlogo:evaluateFisSingletonCentroid FIS inputList 101 0 0 0 0**.

Where the FIS variable contains the entire structure of fuzzy inference systems, the inputList variable is a list that includes the values for the FIS inputs that are required to make a decision.

Other variables are initial configurations that are not recommended to change to unless the researcher is familiar with the process of evaluation of the diffuse logic.

3 Use Cases

There are two different approaches to setting up a fuzzy system. The first consists of an empirical configuration and the second utilizes data mining processes to discover the values of membership functions and rules. In this section, we leverage two agent-based models from the NetLogo model library. Each model was added a FIS so that the agents had their diffuse concepts and their diffuse rules implemented. The objective is to illustrate a possible way of using diffuse inference systems in an agent-based model, with the help of the proposed tool.

3.1 Use Case 1: Empirical Configuration FIS

For this use case, we used the voting model included in the default NetLogo examples library [3]. In this example, we assume that the FIS is configured by a subject expert, using the proposed tool. This model simulates voting distribution through a cellular automaton, each patch decides for "vote," and this can change your "vote" taking into consideration of its eight neighbors.

The model plays the following rules:

- The SETUP button creates an approximately equal but random distribution of blue and green patches. In addition it performs the initial configuration of the FIS by calling the setup-fis method, the.nls file show this method.
- When both switches are off, the central patch changes its color to match the majority vote, but if there is a 4–4 tie, then it does not change.
- If the CHANGE-VOTE-IF-TIED? The switch is on, then in the case of a tie, the central patch will always change its vote.
- If the AWARD-CLOSE-CALLS-TO-LOSER? The switch is on, then if the result is 5–3, the central patch votes with the losing side instead of the winning side.
  The author suggests trying other voting rules, so we added the following rule:
- If the CHANGE-VOTE-BY-PREFERENCE? The switch is on; then the preference is evaluated using the FIS (Fuzzy Inference System), the central patch will change its vote depending on the preference depicted by the FIS. This rule consists of counting all blue preferences and green preferences. These are added to a list and passed as a parameter to the method jt2fisnetlogo: evaluateFisSingletonCentroid who is in charge of evaluating the decision to vote for the FIS configuration.

The FIS was empirically configured, with two inputs and one output. The inputs of the system are BlueVotes and GreenVotes, and the output is BluePreference. The inputs are transformed from crisp data to linguistic values LowBlueVotes, HighBlueVotes, LowGreenVotes, HighGreenVotes by the fuzzify method, then the corresponding output linguistic value is inferred by the inference system, to finally de-fuzzify is applied to get a crisp output data. Figure 10 shows the structure of the FIS for this particular case study.

The FIS represents the preference of all voters that is described by functions membership and rules into the system. The fuzzy system implements the following rules:

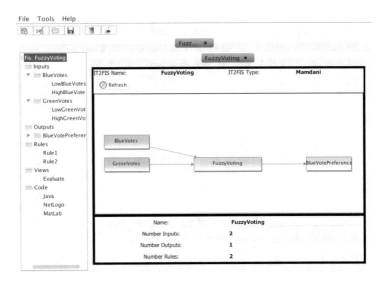

Fig. 10 Fuzzy inference system generated for voting model

- Rule 1: LowBlueVotes and HighGreenVotes then Low-BluePreference.
- Rule 2: HighBlueVotes and LowGreenVotes then HighBluePreference.

Figure 11 shows the results of the model after adding the new rule and executing the simulation.

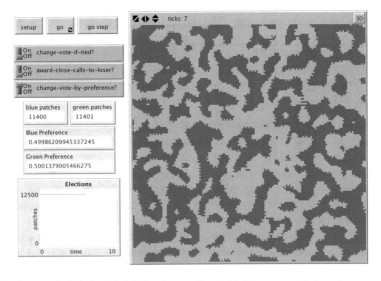

Fig. 11 Case study "Voting Model" implementation in NetLogo screenshot capture

3.2   Use Case 2: Data Mining Configuration FIS

For this use case, we model religious segregation in city of Tijuana. Tijuana is a border city in the northeast of Mexico. Tijuana is a city that has overgrown in recent years as a product of migration. As a consequence, this city has a great diversity of cultures, traditions, and religions [34].

Instituto Nacional de Estadistica y Geografia (INEGI) has provided us with the necessary data for this case study. INEGI is a federal government organization responsible for collecting economic, geographic and socio-demographic data [35]. For our case study, we have considered two different locations.[1] We take into account the 2010 population census in Mexico [35] to select the variables of our model. These variables show below:

1. P15YMAS = Population over 15 years old.
2. P15YMSE = Population over 15 years old without education.
3. GRAPROES = Education.
4. PEA = Working population.
5. PEINAC = Non-working population.
6. PCATOLICA = Catholic population.
7. PNCATOLICA = Non-catholic population.

The next step is to take the data and submit it to a process of data mining, the method chosen was FuzzyCMeans method. We select variables 1–5 as inputs and variables 6–7 outputs. This process can observe in Fig. 12. With the FuzzyCMeans method, we construct a System of Inferential Fuzzy for the Location 187 and 293 for both cases we select 3 clusters. The Fig. 13 show the FIS Location 187 obtained.

After that, we include the FIS generated in the NetLogo Segregation Model. This FIS determines whether an agent is catholic or non-catholic. If the agents do not have enough neighbors of the same religion preference, they move to a nearby patch. To distinguish religious preference, we use a red color for Catholics and green for non-Catholics. The model has the following rules:

- The SETUP button create the initial configuration of the FIS by calling the setup-fis method, this method is found in the.nls file. Choose the religious preference of each agent, red for Catholics and green for non-Catholics. The FIS calculated the religious preference. The values for FIS inputs are calculated randomly.
- DENSITY slider controls the occupancy density of the neighborhood (and thus the total number of agents).
- The %-SIMILAR-WANTED slider controls the percentage of same-color agents that each agent wants among its neighbors. For example, if the slider is \%set at 30, each green agent wants at least 30\% of its neighbors to be green agents.

---

[1]Locations are the terminology used to describe wide geographic areas of the city that are composed of Basic Geo-Statistic Area (BGSA).

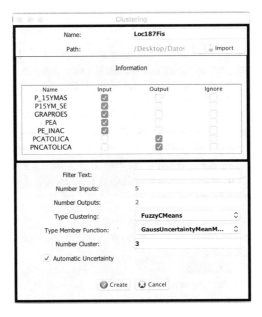

Fig. 12 Screen printing of the clustering process for location 187

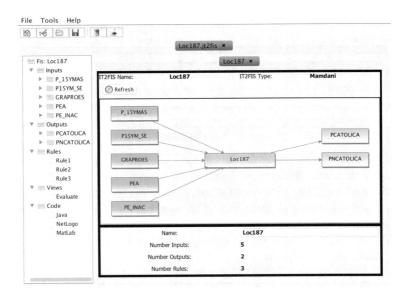

Fig. 13 Fuzzy inference system generated for location 187

- The % SIMILAR monitor shows the average percentage of same-color neighbors for each agent.
- The NUM-UNHAPPY monitor shows the number of unhappy agents, and the % UNHAPPY monitor shows the percent of agents that have fewer same-%color neighbors than they want (and thus want to move). The % SIMILAR and the NUM-UNHAPPY monitors are also plotted.

The importance of creating an FIS with some method of grouping extracting the rules and their configuration of the data is that it becomes a model more attached to the reality. You can also observe the preferences of a population at different levels, either at the level of a locality, colony, country. We use two different simulations, one for location 187 and one for location 283. The Figs. 14 and 15 show the results

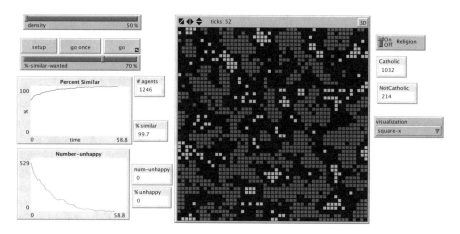

Fig. 14 Religious segregation simulation for location 187

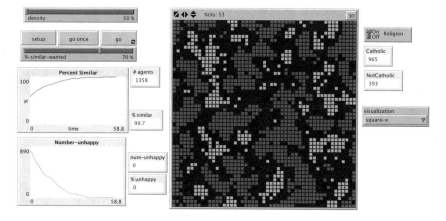

Fig. 15 Religious segregation simulation for location 283

of these simulations. We can see that locality 187 has a greater preference for the Catholic religion than the locality 283.

4 Discussion and Applications

We have leveraged our application to illustrate how to build an FIS from user modeling actions, directly on the visual application, as well as by data mining from real data.

In the first use case, the user must know how to build the FIS configuration step by step, in order to get the desired system. The system will use the experience of the modeler and provide consistent outputs. This kind of system may be the most convenient example for starting to build simple decision-making systems that can be built upon. The advantage is that the modeler can assume all de rules and values and build the system theoretically, so to explore its impact into the full model. The disadvantages are that the modeler must know the purpose of member functions, rules, fuzzify and defuzzify FIS mechanisms, and their meaning in the model.

In the second use case, the user not necessarily must know how to build the FIS, but he can discover the rules and configurations representing hidden information by data mining real data. The system will represent knowledge discovered from the system and will provide consistent outputs with the data, but not necessarily as expected by the modeler. This kind of system could be interesting to add complexity and therefore improve an existing decision-making system. The advantage is that the modeler can discover all de rules and values and build the system realistically to reproduce the result of it into the full model. The disadvantages are that the user depends heavily on the original data (the data must adequately gather and formatted conveniently by the researcher), and the discovered member functions and rules could become complex and therefore difficult to understand by the modeler (the meaning of they into the model could be confused if there are a lot of rules or member functions discovered).

In either case, the modeler can export each generated configuration to the NetLogo model, and build the obtained system by programming, leveraging the extension. The focus can then be shifted to incorporate the FIS, e.g. as a decision-making system, to produce an imitative simulation, without having to worry about member functions, rules, fuzzify and defuzzify FIS mechanisms.

4.1 From Simplistic to Realistic Model

Most of the agents-based simulations are simple agent-based designs that model features from a traditional point of view. Researchers use models to create a standard or example for comparing the study case features in different scenarios. A model could be a representation to show the structure or behavior of an interesting phenomena, e.g. to describe the observable activity in humans, or the aggregate of responses to internal and external stimuli. Sometimes, they use to set up behavior proceeding or derived from reason or based on reasoning programmed by the modeler, many times characterizing a process of selection in which each item of a set has an equal probability of being chosen.

However, models currently require the description of behavior depending upon experience or observation, not necessary using the scientific method or clearly set theory. Moreover, the knowledge communicated or received concerning a particular fact or circumstance should be considered different (unlike or incongruous) or composed of parts of different kinds; having widely dissimilar elements or constituents, rather than same kind or nature. We expect future models based on explanatory reasoning to generate or justify hypotheses, helping inference to the best explanation as a methodology. Realistic models should help researchers in the discovery process, suggesting a new hypothesis that follows a distinctive logical pattern, as a result of computing, rather than both inductive logic and the logic of hypothetical-deductive reasoning.

4.2 Fuzziness and Uncertainty in Agent Behavior

Uncertainty and fuzziness are two critical concepts for representing agent behavior in a given environment. Fuzziness is a central component of behavior, since realistic setting decisions made by the use of heuristics may not represent results from clear rules. However, alternative and possibly competing heuristics are evaluated regularly, potentially leading to significantly different outcomes under relatively similar situations. We may therefore say that decision making in a fuzzy system represents a relative and contextual endeavor, with agents that consider what they perceive and what they know, with an ubiquitous feeling of uncertainty. Because agents always consider some degree of uncertainty perception and knowledge when making decisions, the fuzziness of the system introduces a level of uncertainty to the final choice. Therefore, we can say that the process of decision making for an agent is not only a fuzzy system, but rather a compatibility threshold between what the agents sense and their knowledge, itself also fuzzy.

There continues to be an active discussion in the literature concerning the degrees of uncertainty in a fuzzy system. In the meantime, we can assume that degrees of uncertainty allow us to represent the level of influence that different levels of reality, as expressed in the simulation, especially since agents interact in a social network. We can thus represent agents that qualify the dispersion (or fuzziness) of the data generated by these interactions as contained intrinsically, with corresponding perceptions that can vary, and thus are modeled individually.

4.3 Opportunity Areas for FISs Applications

We identified some opportunities areas where we consider that researchers could use a FIS. For instance, FIS can be utilized mainly as a system able to evaluate a set of inputs with a complex non-linear function for description or prediction. The FIS could represent an environmental multi-variable aspect of the system for example but could be utilized as a Decision-Making System into an agent besides.

Information and communication technologies. In ICT research and technology, FIS could be used as a system profiler, able to evaluate different aspects of users for classification.

Socio-ecological systems. In social sciences modeling, FIS could be utilized as a Decision-Making Systems using agents to describe how individuals made decisions from multiple and complex social, as well as environmental stimuli.

Biological and (bio)medical complexity. In health sciences modeling, FIS could be utilized as a prediction function in disease propagation to risk and contingency management.

Infrastructure, planning, and environment. In urban planning and public policy modeling, FIS could be used as an institution or organization decision-making system that evaluate actions according to a set of public policies and environmental values, e.g. in infrastructure planning [36, 37].

Economics and finance. In businesses and financial modeling, FIS could be used as a system able to evaluate different aspects of complex multi-variable systems to predict prices and market changes.

Cognition and Linguistics. In knowledge management modeling, FIS could be used as a communication/interpretation system able to allow for word-based computation.

5 Conclusions and Future Work

In this work, we demonstrate how to include Type-1 and Type-2 Fuzzy Inference Systems and Data Mining for Agent-Based Simulation in NetLogo. As an application use case, we modified the segregation model and voting model provided in NetLogo example library to show how to develop fuzzy inference systems in NetLogo.

We use the voting model for exemplifying how to use uncertainty and perception on agent-based simulation. The toolkit can apply for developing decision-making system, where individuals choose to donate to a charitable cause, or rent or buy a home. Furthermore, we use the segregation model allows us to exemplify how to use data mining algorithms for discovering the member functions and rules to develop an FIS from a data set. The toolkit introduces in this paper show how to use a real data set for developing a decision system in which individuals decide on whether to Catholic or not Catholic groups preference.

With this toolkit, the researcher can create, discover, explore, and import into NetLogo a Type-1 or Type-2 Fuzzy Inference System using a NetLogo extension to incorporate a machine-learning system as a decision-making system into simulations. FIS can be used to describe of complex systems or for predict systems response to add a decision-making system to agents.

Moreover, a FIS is a Hybrid Intelligent Systems if it could be generated from real data set using clustering algorithms, and it can be used to describe complex systems and predict systems response, adding complexity to an agent's decision-making. This tool provides a least two well-known clustering methods, fuzzy c-means, and subtractive algorithms, in order to produce a Mamdani or Sugeno Type-1 or Type-2 Fuzzy Inference Machine respectively.

Our future work is to upgrade the extension to work properly in the new version NetLogo. We are developing clustering methods to incorporate in the latest version of our toolkit and improve the extraction of the FIS. Our next step is to include neuro-diffuse systems in our tool. With this new feature, we can improve the behavior of agents, since they can add, modify or delete rules in their behavior in a dynamic way. We goal for the future is creating a software agent smart enough for modeling realistic human actors in decision-making systems.

Acknowledgements We thank the MyDCI program of the Division of Graduate Studies and Research, Autonomous University of Baja California, Mexico, the financial support provided by our sponsor CONACYT contract grant number: 257863.

References

1. Bonabeau E (2002) Agent-based modeling: methods and techniques for simulating human systems. PNAS 7280–7287
2. Jennings NR (2001) An agent-based approach for building complex software systems. Commun ACM 44:35–41. https://doi.org/10.1145/367211.367250
3. Wilensky U (1999) NetLogo, 4th ed. Center for Connected Learning and Computer-based Modelling, Northwestern University, Evanston
4. Richiardi M, Leombruni R, Saam NJ, Sonnessa M (2006) A common protocol for agent-based social simulation. J Artif Soc Soc Simul 9:15
5. Mair C, Kadoda G, Lefley M et al (2000) An investigation of machine learning based prediction systems. J Syst Softw 53:23–29. https://doi.org/10.1016/S0164-1212(00)00005-4
6. Zadeh LA (1996) Fuzzy logic = computing with words. IEEE Trans Fuzzy Syst 4:103–111. https://doi.org/10.1109/91.493904
7. Hájek P (1998) Metamathematics of fuzzy logic, 1st ed. https://doi.org/10.1007/978-94-011-5300-3
8. Cintula P, Hájek P, Noguera C (2011) Handbook of mathematical fuzzy logic, vol 1, Petr Cintu. College Publications, London
9. Wang PP, Ruan D, Kerre EE (2007) Fuzzy logic—a spectrum of theoretical & practical issues, 1st ed. https://doi.org/10.1007/978-3-540-71258-9
10. Jang JSR, Sun CT, Mizutani E (1997) Neuro-fuzzy and soft computing: a computational approach to learning and machine intelligence. Prentice Hall, Delhi

11. Ren Q, Baron L, Balazinski M (2006) Type-2 Takagi-Sugeno-Kang fuzzy logic modeling using subtractive clustering. In: NAFIPS 2006–2006 Annual Meeting North American Fuzzy Information Processing Society, pp 120–125

12. Qun R, Pascal B (2017) A highly accurate model-free motion control system with a Mamdani fuzzy feedback controller combined with a TSK fuzzy feed-forward controller. J Intell Robot Syst 86:367–379. https://doi.org/10.1007/s10846-016-0448-7

13. Han J, Kamber M (2006) Data mining: concepts and techniques, 2nd ed. Morgan Kaufmann, Burlington

14. Crows T (1999) Introduction to data mining and knowledge discovery, 3rd ed. Two Crows Corporation

15. Bozkir AS, Sezer EA (2013) FUAT—a fuzzy clustering analysis tool. Expert Syst Appl 40:842–849. https://doi.org/10.1016/j.eswa.2012.05.038

16. Bezdek JC, Ehrlich R, Full W (1984) FCM: the fuzzy c-means clustering algorithm. Comput Geosci 10:191–203. https://doi.org/10.1016/0098-3004(84)90020-7

17. Vaidehi V, Monica S, Mohamed, et al (2008) A prediction system based on fuzzy logic. In: Proceedings of World Congress Engineering Computer Science (WCECS)

18. Bonabeau E (2002) Agent-based modeling: methods and techniques for simulating human systems. Proc Natl Acad Sci 99:7280–7287. https://doi.org/10.1073/pnas.082080899

19. Niazi M, Hussain A (2011) Agent-based computing from multi-agent systems to agent-based models: a visual survey. Scientometrics 89:479. https://doi.org/10.1007/s11192-011-0468-9

20. Niazi MA (2013) Complex adaptive systems modeling: a multidisciplinary roadmap. Complex Adapt Syst Model 1:1. https://doi.org/10.1186/2194-3206-1-1

21. El-Nasr MS, Yen J, Ioerger TR (2000) FLAME-fuzzy logic adaptive model of emotions. Auton Agent Multi Agent Syst 3:219–257. https://doi.org/10.1023/A:1010030809960

22. Ghasem-Aghaee N, Ören T (2003) Towards fuzzy agents with dynamic personality for human behavior simulation. In: Proceedings of 2003 Summer Computer Simulation Conference, pp 3–10

23. Ören T, Ghasem-Aghaee N (2003) Personality representation processable in fuzzy logic for human behavior simulation. In: Proceedings of 2003 Summer Computer Simulation Conference, pp 11–18

24. Rino F, Pezzulo G, Cristiano C (2003) A fuzzy approach to a belief-based trust computation. In: Trust Reputation, Security Theory Practice: AAMAS 2002 International Workshop, Bologna, Italy, 15 July 2002. Selected and Invited Papers. Springer, Berlin and Heidelberg, pp 73–86

25. Ramchurn SD, Jennings NR, Sierra C, Godo L (2004) Devising a trust model for multi-agent interactions using confidence and reputation. Appl Artif Intell 18:833–852. https://doi.org/10.1080/0883951049050904509045

26. Fort H, Pérez N (2005) Economic demography in fuzzy spatial dilemmas and power laws. Eur Phys J B 44:109–113. https://doi.org/10.1140/epjb/e2005-00105-8

27. Neumann M, Braun A, Heinke EM et al (2011) Challenges in modelling social conflicts: grappling with polysemy. J Artif Soc Soc Simul 14(3): 9 . https://doi.org/10.18564/jasss.1818

28. Acheson P, Dagli CH, Kilicay-Ergin NH (2014) Fuzzy decision analysis in negotiation between the system of systems agent and the system agent in an agent-based model. CoRR arXiv preprint arXiv:1402.0029

29. Machálek T, Cimler R, Olševičová K, Danielisová A (2013) Fuzzy methods in land use modeling for archaeology. In: Proceedings of Mathematical Methods in Economics

30. Kim S, Lee K, Cho JK, Kim CO (2011) Agent-based diffusion model for an automobile market with fuzzy TOPSIS-based product adoption process. Expert Syst Appl 38:7270–7276. https://doi.org/10.1016/j.eswa.2010.12.024

31. Lee K, Lee H, Kim CO (2014) Pricing and timing strategies for new product using agent-based simulation of behavioural consumers. J Artif Soc Soc Simul 17:1. https://doi.org/10.18564/jasss.2326

32. Castañón-Puga M, Castro JR, Flores-Parra JM et al (2013) JT2FISA Java Type-2 fuzzy inference systems class library for building object-oriented intelligent applications BT— advances in soft computing and its applications: 12th Mexican International Conference on Artificial Intelligence, MICAI 2013, Mexico City. In: Castro F, Gelbukh A, González M (eds). Springer, Berlin and Heidelberg, pp 204–215

33. Castanon-Puga M, Flores-Parra J-M (2014) JT2FISNetLogo [Online]. http://kiliwa.tij.uabc.mx/projects/jt2fisnetlogo

34. Castanon-Puga M, Jaimes-Martinez R, Castro JR, et al (2012) Multi-dimensional modelling of the religious affiliation preference using Type 2 Neuro-fuzzy system and distributed agency. In: 4th International. Conference on Social Simulation (WCSS 2012)

35. INEGI (2010) Censo de Población y Vivienda 2010. Instituto Nacional de Estadística Geografía e Informática

36. Chang N-B, Parvathinathan G, Breeden JB (2008) Combining GIS with fuzzy multicriteria decision-making for landfill siting in a fast-growing urban region. J Environ Manage 87:139–153. https://doi.org/10.1016/j.jenvman.2007.01.011

37. Sui DZ (1992) A fuzzy GIS modeling approach for Urban land evaluation. Comput Environ Urban Syst 16:101–115. https://doi.org/10.1016/0198-9715(92)90022-J

An Approach to Fuzzy Inference System Based Fuzzy Cognitive Maps

Itzel Barriba, Antonio Rodríguez-Díaz, Juan R. Castro and Mauricio A. Sanchez

Abstract In the search of modeling methodologies for complex systems various attempts have been made, and so far, all have been inadequate in one thing or another leading the pathway open for the next better tool. Fuzzy cognitive maps have been one of such tools, although mainly used for decision making in what-if scenarios, they can also be used to represent complex systems. In this paper, we define an approach of fuzzy inference system based fuzzy cognitive map for modeling dynamic systems, where the complex model is defined by means of fuzzy IF-THEN rules which represent the behavior of the system in an easy to understand format, therefore facilitating a tool for complex system design. Various examples of dynamic systems are shown used as a means to demonstrate the ease of use, design and capability of the proposed approach.

1 Introduction

These days, one of the most important challenges is modeling complex systems (CS). This can range from modeling all things to be considered in a phenomenon, to things which are difficult to represent. CS are sets of interconnected elements that interact between them in order to achieve certain goals. Individual elements on these systems may be easy to define, but as soon as interactions are taken into account, the overall behavior is not easy to handle, and how to disentangle this behavior is not clear, there are no mathematical models to do such thing. Some examples of this type of systems are: human brain, social communities and world economics.

Over the years, it has been noted that, as soon as interactions get to a certain level, emergent properties like nonlinearity, adaptation, feedback loops, etc., start to

I. Barriba (✉) · A. Rodríguez-Díaz · J. R. Castro · M. A. Sanchez
Facultad de Ciencias Quimicas e Ingenieria, Universidad Autonoma de Baja California, Tijuana, Mexico
e-mail: itzel.barriba@uabc.edu.mx

© Springer International Publishing AG, part of Springer Nature 2018
M. A. Sanchez et al. (eds.), *Computer Science and Engineering—Theory and Applications*, Studies in Systems, Decision and Control 143,
https://doi.org/10.1007/978-3-319-74060-7_8

rise. There has been a large number of attempts to model CS, amongst which are complex networks (e.g. biological system [1–3], climate models [4], traffic systems [5]), multi-agent system [6, 7]; social simulations [8–10] and fuzzy cognitive maps (FCM) (e.g. modeling of virtual worlds [11], economic and business [12, 13] medicine [14–17], production system [18–20]). In this paper, since a FCM is considered a useful tool to modeling dynamical system as a decision support and forecast tool, our focus is on FCM.

FCM is a graphical technique based on directed graphs, built using expert knowledge and experience. Nodes represent concepts (which can be objects, places, events, etc.), and edges represent causal relations between those concepts. The behavior of the system can be seen in the directed graph or using the convergence matrix. There are different proposals of FCM besides of traditional as an extension, between them FCM based on rules.

Carvalho et al. [21, 22] proposed Rule Based Cognitive Maps (RB-FCM) where define each concept with several membership functions, and relations are defined by multiple fuzzy rule bases. They created a mechanism called Fuzzy Causal Relations to represent non-monotonic and non-asymmetric causal relations, use a Fuzzy Carry Accumulation (FCA) when two edges affect to a one concept and these exceeded the value of 1, to accumulate the effect.

Another representation is a framework for fuzzy rule-based cognitive maps proposed by Khan [23] where the mapping of the FCM uses an aggregation operator to merge multiple causal inputs, and utilizes a defuzzification to obtain a single fuzzy rule between a cause node to an effect node.

In this paper, we present a new approach to fuzzy cognitive maps with the following contributions:

1. A Fuzzy Inference System is used as the main inference system of the FCM.
2. The FCM behavior is defined by a set of fuzzy rules.
3. Relations among concepts can now be defined by 1 to N relations and N to 1 relations, instead of monotonic only relations.

The rest of this paper is organized as follows: Sects. 2 and 3 review literature of FCM and type 1 Fuzzy Inference Systems, Sect. 4 describes our proposed method for FCM building using fuzzy inference system. Section 5 describes experiments using our proposed method and compares it with other methods, and finally, in Sect. 7 we draw our conclusions.

2 Fuzzy Cognitive Maps

In 1976, Axelrod developed the cognitive maps (CM) [24], to show how decision makers and policy experts make decisions and how to improve such decisions. Of course, this first model was simplistic in the real world, such that causal relations are not only positive or negative but very complex in reality. CM are directed graphs that consist of concepts (nodes) and causal relations (edges) between them, concepts represent some aspects of a system, causal relations can be positive, negative or neutral depending on their effect on those concepts to which they are connected to. These maps are easy to use: an expert defines concepts and relations according to his criteria, then assigns weights as positive or negative (or none if neutral) to all relations depending on how one concept affects another in direct (positive) or inverse (negative) manner.

Based in CM, Kosko proposed the FCM [25] as an extension to CM where relations are fuzzy. Node variables are normalized to a numerical value in the range [0, 1] or [−1, 1]. Also, relations could be a linguistic variable like *little, much,* etc., that is represented by a numerical value in [−1, 1], instead of using positive or negative signs like CM. A simple example is shown in Fig. 1, where nodes are represented by ovals and the causal relationship by arrows.

The FCM also can be represented by an adjacency matrix as shown in Table 1. Causal relations are represented by $e_{ij} = e(C_i, C_j)$, which represents the effect of concept C_i over concept C_j. If $e_{ij} = 1$ then a positive (direct) causality exists, if $e_{ij} = -1$ then a negative (indirect) causality exists, and if $e_{ij} = 0$ then there isn't any causality.

The state of concepts is given in an initial vector as shown in Eq. (1), this vector is multiplied by the adjacency matrix E. The inference of the FCM is given by Eq. (2).

$$C_{i_{(t)}} = \left[C1_{(t)}, \ldots, Cn_{(t)} \right] \tag{1}$$

$$C_j(t+1) = f\left(\sum_{i+1}^{N} e_{ij} C_i(t) \right) \tag{2}$$

where $C_i(t)$ is the value of ith node at the tth iteration, e_{ij} is the edge weight (relationship strength) from the concept C_i to the concept C_j, t is the corresponding iteration, N is the number of concepts, and f is the transformation function.

Because concepts values can only be values between the range of [0, 1] a transfer function is used to normalize these values within the required range. Some

Fig. 1 Sample fuzzy cognitive map

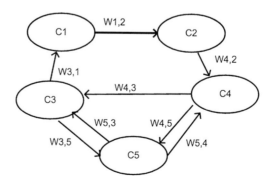

Table 1 Adjacency matrix E of a sample FCM

	C1	C2	C3	C4	C5
C1	0	W1, 2	0	0	0
C2	0	0	0	W2, 4	0
C3	W1, 3	0	0	0	W3, 5
C4	0	0	W4, 3	0	W4, 5
C5	0	0	W5, 3	W5, 4	0

of the transfer functions which are common are: step function, sigmoid function, and hyperbolic tangent function, as shown in Eqs. (3)–(5) respectively. The most utilized one is sigmoid [26].

$$f(C_{it}) = \begin{cases} 0 \ if \ C_{it} \leq 0 \\ 1 \ if \ C_{it} > 1 \end{cases} \tag{3}$$

$$f(C_{it}) = \frac{1}{1 + e^{-\lambda C_{it}}} \tag{4}$$

$$f(C_{it}) = \tanh(\lambda C_{it}) = \frac{e^{\lambda C_{it}} - e^{-\lambda C_{it}}}{e^{\lambda C_{it}} + e^{-\lambda C_{it}}} \tag{5}$$

The stability or inference of a FCM can converge to an attractor (fixed) point or a limit cycle after some iterations. This depends on the system being modeled, if the system is complex or chaotic, the FCM could not reach an attractor, therefore it would be a non-periodic or chaotic attractor.

Some of the advantages of FCMs is that they are easy to use, by simply looking at the FCM one could see the behavior and how concepts are affected by one another by causalities between them, empiric knowledge can also be incorporated [27]. They are also considered a useful prediction tool to decision making. Although there are limitations, relations are only monotonic (i.e. the relations are from one concept to another), and the values of relationships aren't truly fuzzy numbers, instead they use a crisp value [−1, 1], among others. Due to these limitations, some variations or extensions of FCM exist in the literature, e.g. dynamical cognitive networks [28], fuzzy grey cognitive maps [29], dynamic random fuzzy cognitive maps [30], rule-based fuzzy cognitive maps (RBFCM) [21–23, 31–33], etc.

3   Type-1 Fuzzy Inference Systems

A block diagram of Type 1 Fuzzy Logic System (FLS) is showed in Fig. 2, it presents the main elements of the FLS. A FLS maps crisp inputs into crisp outputs. First, the input of a FLS is a crisp value that is converted into a fuzzy set (FS) by a fuzzifier; then, the inference takes these FSs and based on available Rules, maps them to a FS output; finally, the defuzzifier converts the FS output to a crisp value called y.

A type-1 FS, A, is characterized by a membership function $\mu_A(x)$, where $x \in X$, and is defined by $A = \{(x, \mu_A(x)) | x \in X\}$.

Type 1 FSs are described by membership functions that are completely certain, yet represent imprecision or ambiguity. The number of memberships is dependent of the expert, and is associated to linguistic variables whose values are not numbers. An example of linguistic variables for representing *error* can be composed of {*very negative, negative, none, positive, very positive*}. Membership functions can be represented by Gaussian, triangular, g-bell, among other types of membership functions.

A fuzzy rule base consists of a collection of IF-THEN rules, which can be expressed as:

$$R^l : IF\ x_1\ is\ F_1^l\ and \ldots and\ x_p\ is\ F_p^l,\ THEN\ y\ is\ G^l \tag{6}$$

where $l = 1, 2, \ldots, M$, F_1^l and G^l are FSs. These rules represent a relation between input space and output space. R^l is a specific rule, x_p is the input p, F_p^l is a membership function on rule l and input p, y is the output on membership function G^l.

The inference uses the principles of fuzzy logic to combine rules in the format of IF-THEN from a rule base, also uses fuzzy input sets $X_1 \times \cdots \times X_p$ which maps fuzzy output sets in Y. Based on Eq. 5, it is defined as:

$$\mu_{B^l}(y) = \mu_{G^l}(y) \tilde{*} \left\{ \left[\sup_{x_1 \in X_1} \mu_{X_1}(x_1) \tilde{*} \mu_{F_1^l}(x_1) \right] \right. $$
$$\left. \tilde{*} \cdots \tilde{*} \sup_{x_p \in X_p} \mu_{X_p}(x_p) \tilde{*} \mu_{F_p^l}(x_p) \right\}, y \in Y \tag{7}$$

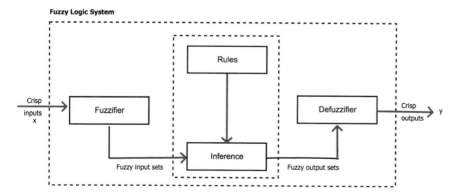

Fig. 2 T1FLS block diagram

The defuzzifier process produces a crisp output from output fuzzy sets. There are many defuzzifiers. For our proposal we only define the centroid defuzzifier. Where the centroid determines the center of gravity (centroid), $\bar{y}$, of B and uses this value as the output of the FLS. From calculus, we know that

$$\bar{y} = \left[\int_S y\mu_B(y)dy\right] \bigg/ \left[\int_S \mu_B(y)dy\right] \tag{8}$$

where S denotes the support of μ_B. Frequently, S is discretized, so that $\bar{y}$ can be approximated by the following formula which uses summations instead of integrations:

$$\bar{y} = \left[\sum_{i=1}^{I} y_i\mu_B(y_i)dy\right] \bigg/ \left[\sum_{i=1}^{I} \mu_B(y_i)dy\right] \tag{9}$$

4 Proposal of Fuzzy Inference System Based Fuzzy Cognitive Maps

A Fuzzy Inference System based Fuzzy Cognitive Maps (FISbFCM) is proposed which can be used to model dynamic systems. FISbFCM is a directed graph with feedback, consisting of concept nodes represented by fuzzy membership functions, and causal relationships between nodes consisting of fuzzy production rules in IF-THEN form. A graphical representation is shown in Fig. 3. The membership functions of a concept can be e.g. *cold, cool, nominal, warm, hot.*, represented through Gaussian, triangular, generalized bell functions, etc., An example of such is shown Fig. 4. These functions represent the state of the concept in a determined step of time.

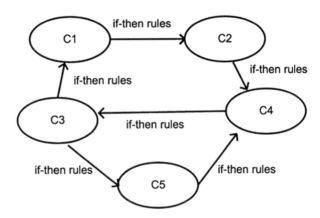

Fig. 3 FISbFCM diagram

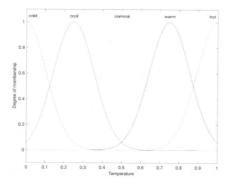

Fig. 4 Membership functions of temperature

The design of a FISbFCM can be based in the experience or knowledge of an expert in the system being modeled. The expert needs to determine the concepts of the system, where each concept should have some membership functions to represent its key variables and the causal between the concepts based on interaction of the concepts. The relationship could be positive or negative.

In FCM the relationships are monotonic, i.e. the interactions are one-to-one, but reality is not always monotonic, therefore using non-monotonic rules is essential to providing a better tool for representing any type system, from simple to complex ones. Using IF-THEN sentences in the rules base to represent the relation in FISbFCM could have more complex relationships like the following:

1 to 1 relation.

If the concept C_i has a directed positive relation to concept C_j as shown in Fig. 5a, the rule could be represented as in Eqs. (10–11).

$$IF\ in1\ is\ low \quad THEN\ out1\ is\ low \tag{10}$$

$$IF\ in1\ is\ high \quad THEN\ out1\ is\ high \tag{11}$$

If the concept C_i has an inverse negative relation to concept C_j as shown in Fig. 5b, the rule could be represented as in Eqs. (12–13).

(a)   **(b)**

Fig. 5 Causal relation of concepts: **a** directed relations of Ci to Cj; **b** inverse relations of Ci to Cj

$$IF\ in1\ is\ low \qquad THEN\ out1\ is\ high \tag{12}$$

$$IF\ in1\ is\ high \qquad THEN\ out1\ is\ low \tag{13}$$

The monotonic relation or 1 to 1 in a FISbFCM are equivalent to a FCM traditional, but if one concept affects to various concepts or one is affected by various, these relations can be produced by the following statements:

1 to N relation.

If two or more concepts like C_j and C_k receive from C_i a directed, inverse or combine relations as shown in Fig. 6a, b, this can be represented as in Eqs. (14–17) respectively.

$$IF\ in1\ is\ low \qquad THEN\ out1\ is\ low \qquad THEN\ out2\ is\ low \tag{14}$$

$$IF\ in1\ is\ low \qquad THEN\ out1\ is\ high \qquad THEN\ out2\ is\ high \tag{15}$$

$$IF\ in1\ is\ low \qquad THEN\ out1\ is\ low \qquad THEN\ out2\ is\ high \tag{16}$$

$$IF\ in1\ is\ low \qquad THEN\ out1\ is\ high \qquad THEN\ out2\ is\ low \tag{17}$$

N to 1 relation.

If the concept C_i receives from two or more concepts a directed, inverse or combine relations for concept C_j and C_k as shown in Fig. 7a, b, this can be represented as in Eqs. (18–21) respectively.

$$IF\ in1\ is\ low\ and\ in2\ is\ low \qquad THEN\ out1\ is\ low \tag{18}$$

$$IF\ in1\ is\ high\ and\ in2\ is\ high \qquad THEN\ out1\ is\ low \tag{19}$$

$$IF\ in1\ is\ low\ and\ in2\ is\ high \qquad THEN\ out1\ is\ low \tag{20}$$

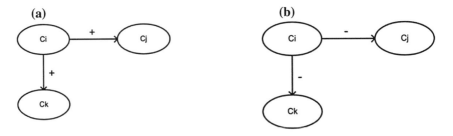

Fig. 6 Relation of concepts: **a** directed relations of 1 to N; **b** inverse relations of 1 to N

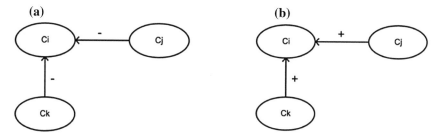

Fig. 7 Relation of concepts: **a** directed relations of N to 1; **b** inverse relations of N to 1

Fig. 8 Diagram of FISbFCM model

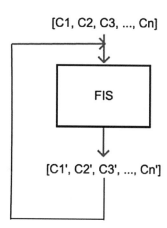

$$IF\ in1\ is\ low\ and\ in2\ high\quad THEN\ out1\ is\ high \tag{21}$$

The proposed FISbFCM is inferred by a single FIS, where the sum of all rules is processed at the same time, therefore simultaneously obtaining the updated values for all concepts. As shown in Fig. 8, the proposed inference is reduced to only a vector of concept values $[C1, C2, C3, \ldots, Cn]$, which is inferred by the formed FIS and a vector of updated values $[C1', C2', C3', \ldots, Cn']$ is obtained, such that these values can be once again be entered to the FIS for obtaining additional steps in the FISbFCM which ultimately carry out the simulation of the system being modeled.

5 Experimental Results and Discussion

Various experiments were performed with FISbFCM to demonstrate the benefits. The first example is a design based in a model prey-predator, has three concepts: grass, sheep (prey) and wolf (predator) as shown in Fig. 9. Where the predator feeds on prey, and the prey feeds on grass. This example is a simple representation

Fig. 9 FISbFCM of
prey-predator model

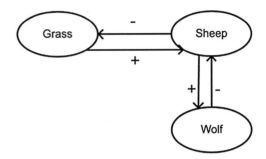

Table 2 Rule base
FISbFCM of prey-predator
model

Rule	Inputs			Output		
	Grass	Sheep	Wolf	Grass	Sheep	Wolf
1	None		A lot		None	
2	Little		Much		Little	
3	Much		Little		Much	
4	A lot		None		A lot	
5		None		A lot		None
6		Little		Much		Little
7		Much		Little		Much
8		A lot		None		A lot

of traditional FCM with positive and negative relationships, to model this with FISbFCM a rule-base as shown in Table 2 was designed.

Experiments were carried out using Gaussian membership functions. Results are shown in Fig. 10a–d, using different vectors of initial values and transfer functions to test the diversity offered by both the FISbFCM and transfer functions.

This first example of the prey-predator dynamic system can be defined by the set of rules as shown in Table 2, where linguistic variables *None, Little, Much*, and *A lot* were used in order to create a granular definition of the desired behavior of causality between concepts. And all knowledge pertaining to the overall behavior of the system is easily interpreted, where each rule in itself gives a piece of information or all rules tell the whole story. In contrast to traditional FCMs where only a single causal relationship at a time can be interpreted, not giving the entire scope at once; this being of the main contributions of the proposed FISbFCM. In relation to Fig. 10a–c denote a similar behavior where the only change is the weights of each concept. Yet an interesting simulation is shown in Fig. 10d, where a cyclical behavior is obtained, such that the expected conduct of all three concepts happen similar as to one would expect. This shows that the proposed technique can be configured in such a way that expected behaviors can be achieved, giving ways for better simulations of complex systems.

The second example is a design based in a model extracted from [34], it is a socio-economic model with only positive and negative causal relationships as

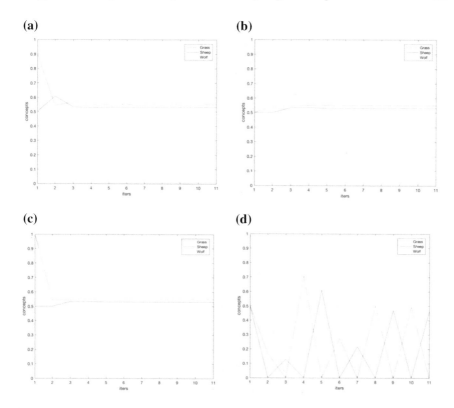

Fig. 10 FISbFCM results for prey-predator system: **a** with initial vector [0.9 0.5 0.5] using sigmoid transfer function; **b** with initial vector [0.5 0.5 1] using sigmoid transfer function; **c** with initial vector [1 0.5 1] using sigmoid transfer function; **d** and with initial vector [0.1 0.5 1] using hyperbolic tangent transfer function

shown in Fig. 11, composed of five concepts: Population, Crime, Economic condition, Poverty and Unemployment. To model this with FISbFCM a rule-base was design with 24 rules, three for each relation.

Experiments were first performed as a traditional FCM, results are shown in Fig. 12a. Then, they were performed using the proposed FISbFCM as shown in Fig. 12b, using different initial vectors and transfer functions, as defined in Fig. 12.

In this example, each concept can be defined by the set of rules selected by an expert. Here, linguistic variables where *Low, Medium* and *High* to create a better definition and try to obtain the desired behavior. In Fig. 12, both models achieve a similar behavior, the map reached an attractor after few interactions.

The last example is a model extracted from [27], the FCM represents city health issues, it has seven concepts and causal relationships have weight appointed by numerical values as shown in Fig. 13, in the domain of [−1, 1]. Concepts are: *Number people in city (N1), Migration into city (N2), Modernization (N3), Amount of garbage (N4), Number of diseases (N5), Sanitation facilities (N6)* and *Bacteria*

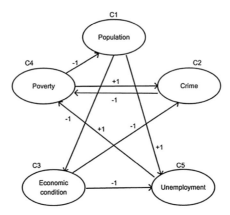

Fig. 11 Diagram of socio-economic modeled with traditional FCM

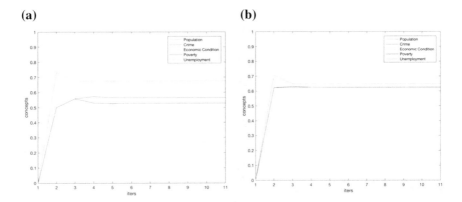

Fig. 12 Results for socio-economic system: **a** with initial vector [1 0 0 0 0] using traditional FCM with sigmoid transfer function; **b** FISbFCM using initial vector [1 0 0 0 0] with sigmoid transfer

per area (N7). This model was created by a rule base consisting of 21 rules, in contrast to the previous example using monotonic relations, since one concept directly affects two or more concepts, therefore the relations 1 to N can be used. Then, IF-THEN statements will be designed as one antecedent and two or more consequents, because of this property rules were reduced to 21 instead of 30.

This experiment was performed using Gaussian and Triangular membership functions. Results are shown in Fig. 14a, b, using a vector of initial values and transfer function to test the diversity of FISbFCM and traditional FCM.

The proposed FISbFCM system can be defined by a set of rules, where linguistic variables such as *Little*, *Medium* and *Much* were used to create each concept, as well as the relationships for all concepts according to weights of each edge. The general behavior is easily interpreted just by visually assessing the rule base. In relation to the concept *Modernization (N3)* can we see that this concept directly

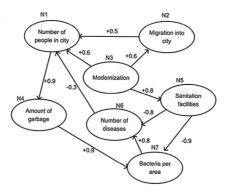

Fig. 13 FCM model of city health issues

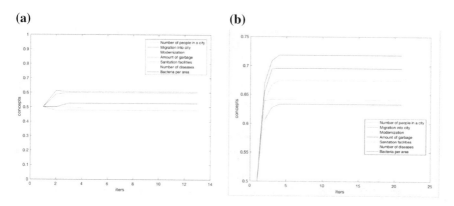

Fig. 14 Results of health issues of a city model: **a** traditional FCM with initial vector [0.5 0.5 0.5 0.5 0.5 0.5 0.5]; **b** FISbFCM using initial vector [0.5 0.5 0.5 0.5 0.5 0.5 0.5] with sigmoid transfer

affects in a positive manner the concepts: *Number of people in city (N1), Migration into city (N2) and Sanitation facilities (N5).* An example of this rulebase is showed in the following:

Rule 1 : *IF N3 is **little** THEN N1 is **little** and N2 is **little** and N5 is **little***

Rule 2 : *IF N3 is **medium** THEN N1 is **medium** and N2 is **medium** and N5 is **medium***

Rule 3 : *IF N3 is **high** THEN N1 is **high** and N2 is **high** and N5 is **high***

In Fig. 14a a simulation can be appreciated using a traditional FCM design, where an attractor is reached after a few interactions, this is shown just to observe a simple comparison between the two simulations. Whereas the simulation of the proposed FISbFCM is appreciated in Fig. 14b, where also an attractor is reached within few iterations. The behavior is very similar for both cases, showing that the proposed FISbFCM can also behave like traditional FCMs.

6   Conclusions

Modeling complex systems had been difficult to represented, using the FISbFCM could be a useful tool to represented since relationships can be better represented with rule base. The status of the concepts is calculated with feedback, using the FIS again, the process is iterative. With the experiments could be proved the behavior of the FISbFCM, the findings were that can be use a to modeling scenarios complex, models with many relationships are able to considerably reduce the rules using different relationships N to 1, 1 to N. The behavior depends on how the model was designed, which is based on the expert, since the model can converge in few iterations or reach a limit cycle. Also modeling existing examples of FCM can see similar behaviors but using rule base, can be obtain more realistic models, adopting the proposed technique.

Acknowledgements We thank the MyDCI program of the Division of Graduate Studies and Research and Universidad Autonoma de Baja California, financial support provided by our sponsor CONACYT contract grant number: 345608.

References

1. Chasman D, Fotuhi Siahpirani A, Roy S (2016) Network-based approaches for analysis of complex biological systems. Curr Opin Biotechnol 39:157–166. https://doi.org/10.1016/j.copbio.2016.04.007
2. Sayama H, Pestov I, Schmidt J et al (2013) Modeling complex systems with adaptive networks. Comput Math with Appl 65:1645–1664. https://doi.org/10.1016/j.camwa.2012.12.005
3. Bader AA, Sherif G, Noah O et al (2017) Modeling and analysis of modular structure in diverse biological networks. J Theor Biol 422:18–30. https://doi.org/10.1016/j.jtbi.2017.04.005
4. Jacobson MJ, Markauskaite L, Portolese A et al (2016) Designs for learning about climate change as a complex system. Learn Instr 52:1–14. https://doi.org/10.1016/j.learninstruc.2017.03.007
5. Yan Y, Zhang S, Tang J, Wang X (2017) Understanding characteristics in multivariate traffic flow time series from complex network structure. Phys A Stat Mech its Appl 477:149–160. https://doi.org/10.1016/j.physa.2017.02.040
6. Leitão P, Barbosa J, Trentesaux D (2012) Bio-inspired multi-agent systems for reconfigurable manufacturing systems. Eng Appl Artif Intell 25:934–944. https://doi.org/10.1016/j.engappai.2011.09.025
7. Ilie S, Bădică C (2013) Multi-agent approach to distributed ant colony optimization. Sci Comput Program 78:762–774. https://doi.org/10.1016/j.scico.2011.09.001
8. Puga-Gonzalez I, Sueur C (2017) Emergence of complex social networks from spatial structure and rules of thumb: a modelling approach. Ecol Complex 31:189–200. https://doi.org/10.1016/j.ecocom.2017.07.004
9. Raducha T, Gubiec T (2017) Coevolving complex networks in the model of social interactions. Phys A Stat Mech Appl 471:427–435. https://doi.org/10.1016/j.physa.2016.12.079
10. de Arruda HF, Silva FN, Costa L da F, Amancio DR (2017) Knowledge acquisition: a Complex networks approach. Inf Sci (Ny) 421:154–166. https://doi.org/10.1016/j.ins.2017.08.091

11. Dickerson JA, Kosko B (1993) Virtual worlds as fuzzy cognitive maps. In: Proceedings of IEEE virtual reality annual international symposium, pp 173–189. https://doi.org/10.1109/VRAIS.1993.380742
12. Glykas M (2013) Fuzzy cognitive strategic maps in business process performance measurement. Expert Syst Appl 40:1–14. https://doi.org/10.1016/j.eswa.2012.01.078
13. Groumpos PP (2015) Modelling business and management systems using fuzzy cognitive maps: a critical overview. IFAC-PapersOnLine 48:207–212. https://doi.org/10.1016/j.ifacol.2015.12.084
14. Kang J, Zhang J, Gao J (2016) Improving performance evaluation of health, safety and environment management system by combining fuzzy cognitive maps and relative degree analysis. Saf Sci 87:92–100. https://doi.org/10.1016/j.ssci.2016.03.023
15. Douali N, Papageorgiou EI (2011) Case based fuzzy cognitive maps (CBFCM) : new method for medical reasoning. IEEE Int Conf Fuzzy Syst 844–850
16. Amirkhani A, Papageorgiou EI, Mohseni A, Mosavi MR (2017) A review of fuzzy cognitive maps in medicine: taxonomy, methods, and applications. Comput Methods Programs Biomed 142:129–145. https://doi.org/10.1016/j.cmpb.2017.02.021
17. Papageorgiou EI, Subramanian J, Karmegam A, Papandrianos N (2015) A risk management model for familial breast cancer: a new application using fuzzy cognitive map method. Comput Methods Programs Biomed 122:123–135. https://doi.org/10.1016/j.cmpb.2015.07.003
18. Peláez EC, Bowles JB (1996) Using fuzzy cognitive maps as a system model for failure modes and effects analysis. Inf Sci (Ny) 88:177–199. https://doi.org/10.1016/0020-0255(95)00161-1
19. Stylios CD, Groumpos PP (1999) Mathematical formulation of fuzzy cognitive maps. In: Proceedings of 7th mediterranean conference on control and automation (MED99), Haifa, pp 2251–2261
20. Mendonça M, Angelico B, Arruda LVR, Neves F (2013) A dynamic fuzzy cognitive map applied to chemical process supervision. Eng Appl Artif Intell 26:1199–1210. https://doi.org/10.1016/j.engappai.2012.11.007
21. Carvalho JP, Tomé J (2000) Rule based fuzzy cognitive maps–qualitative systems dynamics. In: Proceedings of 19th international conference of the North American. Fuzzy information processing society NAFIPS2000, Atlanta, pp 407–411
22. Carvalho JP, Tomé JA (1999) Rule based fuzzy cognitive maps-fuzzy causal relations. Comput Intell Model Control Autom Evol Comput Fuzzy Log Intell Control Knowl Acquis Inf Retrieval, IOS Press
23. Khan MS, Khor SW (2004) A framework for fuzzy rule-based cognitive maps. Pricai 2004. Trends Artif Intell Proc 3157:454–463
24. Axelrod R (1976) Structure of Decision: the cognitive maps of political elites. Princeton University Press, New Jersey
25. Kosko B (1986) Fuzzy Cognitive maps. Int J Man-Mach Stud 24:65–75. https://doi.org/10.1007/978-3-642-03220-2
26. Bueno S, Salmeron JL (2009) Benchmarking main activation functions in fuzzy cognitive maps. Expert Syst Appl 36:5221–5229. https://doi.org/10.1016/j.eswa.2008.06.072
27. Stach W, Kurgan L, Pedrycz W (2005) A survey of fuzzy cognitive map learning methods. Issues Soft Comput
28. Miao Y, Liu Z-Q, Siew CK, Miao CY (2001) Dynamical cognitive network—an extension of fuzzy cognitive map. IEEE Trans Fuzzy Syst 9:760–770. https://doi.org/10.1109/91.963762
29. Salmeron JL (2010) Modelling grey uncertainty with fuzzy grey cognitive maps. Expert Syst Appl 37:7581–7588. https://doi.org/10.1016/j.eswa.2010.04.085
30. Aguilar J (2004) Dynamic random fuzzy cognitive maps. Comput y Sist 7:260–270
31. Carvalho JP, Tomé JA (2001) Rule based fuzzy cognitive maps expressing time in qualitative system dynamics. In: 10th IEEE International conference on fuzzy systems (Cat No. 01CH37297). https://doi.org/10.1109/FUZZ.2001.1007303
32. Carvalho JP, Tomé JA (2002) Issues on the stability of fuzzy cognitive maps and rule-based fuzzy cognitive maps. In: Proceedings of NAFIPS-FLINT 2002 annual meeting of the North

American fuzzy information processings society (Cat No. 02TH8622), pp 105–110. https://doi.org/10.1109/NAFIPS.2002.1018038

33. Carvalho JP, Tomé JA (1999) Fuzzy mechanisms for causal relations. In: Proceedings of the 8th international fuzzy systems association world congress. IFSA, Taiwan

34. Kandasamy DWBV, Smarandache F (2003) Fuzzy cognitive maps and neutrosophic cognitive maps, p 212

Detecting Epilepsy in EEG Signals Using Time, Frequency and Time-Frequency Domain Features

D. E. Hernández, L. Trujillo, E. Z-Flores, O. M. Villanueva
and O. Romo-Fewell

Abstract Seizures caused by epilepsy are unprovoked, they disrupt the mantel activity of the patient and impair their normal motor and sensorial functions, endangering the patient's well being. Exploiting today's technology it is possible toe create automatic systems to monitor and evaluate patients. An area of special interest is the automatic analysis of EEG signals. This paper presents extensive analysis of feature extraction and classification methods that have reported good results in other EEG based problems. Several methods are detailed to extract 52 features from the time, frequency and time-frequency domains in order to characterize the EEG signals. Additionally, 10 different classification models, together with a feature selection method, are implemented using these features to identify if a signal corresponds to an epileptic state. The experiments were performed using the standard BONN and the proposed method achieve results comparable to those in the state-of-the-art for the three and four classes problems.

1 Introduction

We are leaving in a new era of biomedical research, with the rapid growth and accessibility of sophisticated sensing and data processing technologies it is now possible to develop systems that can automatically monitor and evaluate patients health sate. One area that has seen a substantial increase in multidisciplinary work is

D. E. Hernández (✉) · L. Trujillo · E. Z-Flores
TecNM—Instituto Tecnológico de Tijuana, Blvd. Industrial y Av. ITR Tijuana S/N,
Mesa Otay, 22500 Tijuana, Baja California, Mexico
e-mail: daniel.hernandezm@tectijuana.edu.mx

O. M. Villanueva · O. Romo-Fewell
Neuroaplicaciones y Tecnologías S.A.P.I. de C.V., Blvd. Diaz Ordaz 12415, OF. M5-7,
Fracc. El Paraiso, Tijuana, Baja California, Mexico

© Springer International Publishing AG, part of Springer Nature 2018
M. A. Sanchez et al. (eds.), *Computer Science and Engineering—Theory
and Applications*, Studies in Systems, Decision and Control 143,
https://doi.org/10.1007/978-3-319-74060-7_9

the automatic analysis of signals originating from the human brain. The large variety of signal processing, pattern recognition and machine learning paradigms has allowed researchers to propose systems that can learn to detect and classify pathologies [1–3], mental states [4] and even emotions [5, 6].

This work focuses on the automatic detection of epileptic seizures, a neurological disorder where a group of neurons is activated in an abnormal, excessive and synchronized manner [7]. Fairly recent statistics estimate that around 65 million people are living with epilepsy world wide [8]. About 70% of the cases can be controlled by medication, but it is not without side effects [9]. Therefore, a large amount of research has been devoted towards the development of automated systems that can aid in the diagnosis, monitoring and treatment of the disorder. Among the sensing technologies available, the electroencephalogram (EEG) is particularly of interest since it is a relatively low cost, non-invasive and flexible method, but other approaches are also available [1, 2].

An expert physician can be trained to identify the different stages of an epileptic seizure, but automatic methods must solve a complex signal processing and pattern recognition problem given the complexity and nature of the EEG signals, and in general the complexity of the human brain. To solve this task, one approach is to pose a supervised learning problem, and develop a machine learning pipeline that includes: feature extraction, feature selection, training and online classification. In this contribution, we evaluate a broad set of features, considering the time domain, frequency domain and time-frequency domain. Moreover, we also apply a feature selection stage and evaluate a diverse set of popular classifiers. Our work is inspired by [6], that also studied an EEG analysis task (emotion recognition). This broad evaluation can provide a comprehensive overview to new researchers in this field of how to pose and solve the automatic epilepsy detection problem. Our intention will not be to go deep into mathematical formalism's or technical details of the employed algorithms. We will mostly steer the interested reader towards the original works for such information, and we will heavily rely on community standard libraries for machine learning tasks. Instead, the focus will be in clearly explaining how to apply, evaluate and analyze the performance of these methods on this unquestionably difficult task. With any luck, this will encourage engineers, graduate students and young researchers to join the many researchers working in this worthwhile domain.

The remainder of the chapter proceeds as follows. Section 2 will provide relevant background regarding epilepsy and the EEG signals produced by patients of this disorder, and will quickly survey notable and recent related literature. Afterward, Sect. 3 presents the methods used in this study, regarding feature extraction, feature selection and classification. The dataset for our case study, experimental setup and evaluation procedures are outlined in Sect. 4. Experimental results are presented in Sect. 5 along with a detailed discussion. Finally, concluding remarks are presented in Sect. 6.

2 Background

In this section we present some relevant background to our current work. Firstly, we provide a deeper description of epileptic seizures and how they are recorded through EEG recordings. Secondly, we provide a literature review of automatic methods that have been proposed for the detection of epilepsy seizures.

2.1 Epileptic States

Seizures caused by epilepsy are unprovoked, they disrupt the mantel activity of the patient and impair their normal motor and sensorial functions. There are five main stages during and epileptic seizure [10], these are the following: the Basal state; the Pre-Ictal state; the Ictal state; the Post-Ictal state; and the Inter-Ictal state. Each state induces different symptoms from the patient, and the EEG recordings differ accordingly. The Basal state corresponds overall to normal brain activity, and the EEG signals are usually characterized by a high frequency and relatively low amplitude. During the Pre-Ictal state some of the most notable seizure symptoms begin to appear, while the EEG signals will begin to exhibit spikes and recruiting rhythms [11]. However, the most evident symptoms of the seizure appear during the Ictal state, with the EEG signal increasing in amplitude and displays a dominant low frequency rhythm. Afterward, the Post-Ictal state refers to the span of time when an altered state of consciousness exists, this is after the main symptoms of the seizure have dissipated. Finally, the Inter-Ictal state refers to the period of time between seizures. In this chapter, we focus on detecting the onset of the Ictal state, as many works reviewed below have done.

2.2 Related Work

There have been many works devoted to the automatic classification of epileptic seizures, a relatively recent survey is presented by [12]. For comparative purposes researchers often rely on benchmark dataset, a popular one as been the Bonn dataset [13], used extensively over the last fifteen years [3, 12].

Most proposed systems pose a supervised learning approach, where a dataset of human labeled examples are used to train a machine learning or pattern recognition algorithm. Once reliable and human labeled data is obtained and validated by domain experts, then machine learning researches usually implement the following five stage process: (1) signal pre-processing; (2) feature extraction; (3) feature selection; (4) building a generative or discriminative model (a process also called learning or training); and (5) validation of the trained model. Before discussing specific examples, some comments on these processes are useful. First, the first stage will not be heavily discussed in this work, but it is mostly based on well-known signal processing procedures, such as (band-pass) filtering, denoising and/or sub-sampling. Second, the feature selection process is not always necessary, particularly if we are dealing with a relatively small set of features. However, if the feature space is large then many learning algorithms (such as Support Vector Machines) can find it difficult to build effective models. On the other hand, many learning algorithms (Decision Trees, Random Forests, Genetic Programming) have built-in, but implicit, feature selection capabilities. Finally, the second process might be the most difficult one to solve. Correctly defining the correct feature space to consider requires domain expertise along with basic machine learning and signal processing knowledge. In essence, feature extraction maps the input signal or phenomenon of interest (in this case EEGs) to another space, where it is expected that automatic learning can be carried out.

Focusing on extracting features from EEG signals, these can be analyzed based on the information they encode in the time domain, the frequency domain, or both. For instance, in the time domain [14] proposed an improved Dynamic Principal Component Analysis (DPCA) by means of a non-overlapping moving window, and [15] uses the Hilbert Transform (HT) as a method for a time to time-domain transformation. Much simpler statistical features were also effectively used in automatic epilepsy analysis [1, 3]. Such relatively simple features can be exploited, when they are combined with more complex features [3] or when they are used with highly non-linear learning methods [1].

However, probably the most widely used approaches use time-frequency analysis using either wavelet decomposition [3, 16–20] or the Short Time Fourier Transform (STFT) [21]. Other examples include the work by [22] that use the empirical mode decomposition or the work by [23] that approximate the Lyapunov exponent to measure the presence of chaos in the EEG signals. Another effective feature extraction method is the matching pursuit algorithm [2, 3, 10, 24–26], a family of greedy algorithms that compute the best nonlinear approximation of a signal.

3   Applied Methods

In this work, we will not directly follow recent works on automatic epilepsy detection. Instead, we look to a related problem of EEG analysis for inspiration, that of emotion recognition. While obviously presenting a problem-specific subset of challenges (regarding the subjectivity of emotional states), most of the issues from a signal processing and machine learning perspective very much overlap between both tasks. In particular, we take inspiration from the work of [6], where a broad evaluation of many types of features are considered, from the time, frequency and time-frequency domain. Given the size of the feature space, we will also employ a feature selection algorithm, show to be effective in learning EEG patterns. Moreover, we evaluate a large set of classifiers that are popular in the machine learning community.

3.1 Feature Extraction and Selection

This section focuses on describing a wide set of features that have been implemented for describing EEG signals on other classifications problems, such as emotion recognition. In general the features are classified based on their domain: time, frequency and time-frequency. In general, the features are extracted from the captures signal of a single electrode; nevertheless, there is a couple of features that combine several electrodes which were also implemented in this work. Following the notation described in [6]: $\xi(t) \in R^T$ denotes the vector containing the time series from a single electrode, T is the number of samples in ξ. The time derivative is identified by $\dot{\xi}(t)$. A feature of $\xi(t)$ is marked as x, the matrix $X = [x_1, .., x_F]$ contains all the features from all the samples, x_i is the vector of a single feature and F is the number of features.

3.1.1 Time Domain

Since the EEG signals are not periodic, time domain features are not predominant in EEG analysis. Nevertheless, there are several approaches that implement such features for characterizing EEG signals for classification. The first set of features from this domain implemented in this work are the statistics of the signal, the features are defined as follows:

- Power: $P_\xi = \frac{1}{T}\sum_{-\infty}^{\infty} |\xi(t)|^2$
- Mean: $\mu_\xi = \frac{1}{T}\sum_{t-1}^{T} \xi(t)$
- Standard Deviation: $\sigma_\xi = \sqrt[a]{\frac{1}{T}\sum_{t-1}^{T} (\xi(t) - \mu_\xi)^2}$
- 1st difference: $\delta_\xi = \frac{1}{T-1}\sum_{t-1}^{T-1} |\xi(t+1) - \xi(t)|$
- Normalized 1st difference: $\bar{\delta} = \dfrac{\delta_\xi}{\sigma_\xi}$
- 2nd difference: $\gamma_\xi = \frac{1}{T-2}\sum_{t-1}^{T-2} |\xi(t+2) - \xi(t)|$
- Normalized 2nd difference: $\bar{\gamma}_\xi = \dfrac{\gamma_\xi}{\sigma_\xi}$

The P_ξ feature measures the signal's strength or consumed energy per unit of time. The μ_ξ and σ_ξ are the statistical moments of the signal. The 1st and 2nd differences describe how the signal changes over time. Note that $\bar{\delta}_\xi$ is also known as the Normalized Length Density, this feature quantifies self-similarities within the EEG signal [6].

Other time domain features implemented in this research are called Hjorth features [27], composed of three metrics: Activity, Mobility and Complexity. The Activity represents the variance of the signal is defined as:

$$A_\xi = \frac{\sum_{t=1}^{T} (\xi(t) - \mu_\xi)^2}{T}$$

The Mobility is define by the standard deviation of the sole of the signal, using as reference the standard deviation of the amplitude and is expressed as a ratio by time unit. $M_\xi = \sqrt{\frac{var(\dot{\xi}(t))}{var(\xi(t))}}$

Finally, the Complexity measures the signal's variation using a smooth curve as reference. $C_\xi = \frac{M(\dot{\xi}(t))}{M(\xi(t))}$

Another time based feature for EEG analysis is the Non-Stationary Index (NSI) proposed by Hausdorff et al. [28]. The NSI is based on the variation of segments average over time. To obtain this value, the signal ξ is divided into small segments and their μ_i is calculated. The NSI is defined as the standard deviation of the segments' means. Hence, the higher the value, the signal is considered "less stationary".

The last time domain features implemented in this work are the Higher Order Crossings (HOC) proposed by Petrantonakis and Hadjileontiadis [29]. This measure describes the oscillatory pattern of a signal. In order to extract these features, a sequence of high-pass filters is applied over a zero-mean time series $Z(t)$ as follows: $\Im_k\{Z(t)\} = \nabla^{k-1}Z(t)$, where ∇^k is the backward difference operator with order $k = 1, .., 10$. The resulting D_k features are defined by counting the sign changes on the processed signal $\Im_k\{Z(t)\}$.

3.1.2 Frequency Domain

Fourier analysis is one of the basic tools for signal processing. Hence it is widely used for describing EEG signals with different purposes. Some of the most popular features in this domain are power features extracted from different frequency bands. These features are calculated using the power spectrum or Band Power $P(u)$ of $\xi(t)$ defined by $P(u) = |F(u)|^2$ where the magnitude $|F(u)|$ corresponds to the Fourier Spectrum of $\xi(t)$. The frequency bands in which the EEG signals are decomposed for these features are given in Table 1, note that the bands' ranges vary slightly between studies.

These particular features assume stationarity of the signal. This work implements the extraction process described in [6], using the Short Time Fourier Transform (STFT) with Hamming window of one second with no overlap. The features obtained from the resulting segmentcd representation of the signal are: mean power, minimum, maximum and variance from all segments of the signal. These features are extracted for each band. An additional feature is calculated using the β and α bands mean power ratio β/α. These process results in 25 features from a single channel.

Table 1 Typical frequency band ranges for EEG signals [31]

Bandwidth (Hz)	Band Name
1–4	δ
4–8	θ
8–10	slow α
8–12	α
12–30	β
30–64	γ

3.1.3 Time-Frequency Domain

The last kind of features implemented in this research are the from the Time-Frequency domain. An important advantage of this kind of features is they bring additional information by considering dynamical changes on non-stationary signals. The extraction process of these features is based on the Discrete Wavelet Transform (DWT), which is an extension of the STFT. The DWT decomposes the signal in several approximation (A) and detail (D) coefficients levels corresponding to different frequency bands, while preserving the signal's time information. The decomposition level for the frequency bands are listed in Table 2. In this particular case, the Daubechies 4 wavelet at 512 Hz is implemented for processing the EEG segments. Once DWT is complete, the Root Mean Square (RMS) and Recursive Energy Efficiency (REE) features are calculated as follows:

$$RMS(j) = \sqrt{\frac{\sum_{i=1}^{j} \sum_{n_i} D_i(n)^2}{\sum_{i=1}^{j} n_i}}$$

where D_i are the detail coefficients, n_i is the number of D_i at the i th level, and j represents the number of levels. Then, the REE feature is evaluated for each frequency band.

$$REE = \frac{E_{band}}{E_{total-3b}},$$

where E_{band} is the energy of a given band, and the total energy of the frequency bands corresponds to $E_{total} - 3b = E_\alpha + E_\beta + E_\gamma$. The energy of a band is calculated as $E_{band} = \sqrt{\sum_{n_i} D_i^2}$

Table 2 Frequency bands decomposition level for the DWT for EEG signals [31]

Band Name	Decomposition level
δ	A6
θ	D6
slow α	D5 (8–16 Hz)
α	D5 (8–16 Hz)
β	D4
γ	D3

This last feature is obtained for each of the frequency bands, resulting in six features from the Time-Frequency domain.

The features described in this section amount to 52 features that can be used to describe a given EEG signal. Nevertheless, trying to classify inputs with such a high dimensionality can prove to be a complex task. A filter feature selection method was implemented to reduce the feature space. One of the advantages of filter methods over the wrapper techniques is the reduced computational cost.

3.1.4 ReliefF

ReliefF is a feature selection algorithm of the filter kind. It uses a subsample of all the signal instances in order to establish weights over the features that represent their capacity to discriminate between two classes. The algorithm estimates a quality weight $W(i)$ for each feature x_i, $i = 1, .., F$ by calculating the difference to the nearest hit x_H and the nearest miss x_M for m randomly selected features vectors.

$$W(i) = W(i) - diff(i, x_k, x_H)/m + diff(i, x_k, x_M)/m$$

This weight adjustment is an iterative process. Note that this method also considers the multi-class scenario by searching for n nearest instances from each class weighted by class prior probability [3].

3.2 Classification Methods

The "No free lunch" theorem states that there is no single model that works best for every problem. The concept of the best model exists for a single problem, but this condition may not hold for another problem. For this reason, it is fairly common in machine learning problems to test several models and find one that works best for a particular problem. In this paper 10 different classification models are implemented for detecting epileptic formations in an EEG signal. The method used are reported to achieve good classification results in EEG based emotion recognition [6, 30].

The first model used in this work is the proximity based classifier known as k-nearest neighbors (k-NN). The Naïve Bayes classifier that implements Bayes decision theory to find the most probable class for a new observation is also part of the tested models.

Additionally, some linear classifiers were implemented. The major advantage of these models is their simplicity and computational attractiveness. The selected models where the Perceptron, Linear Discriminant Analysis (LDA) and a Support Vector Machine (SVM) with a linear kernel.

Besides the previously mentioned algorithms, five non linear methods where implemented to classify the EEG signals. First the Quadratic Discriminant Analysis (QDA), which is an extension of the LDA model that does not assumes the classes have the same shape. Then, a Multi-Layer Perceptron (MLP) was implemented; this model is a generalization of the Perceptron that includes several hidden layers to define a non linear combination of the input data. The last selected methods were the Decision Tree and the Random Forest models. According to [31] the RF and SVM are the classification methods that achieve better results in most real world problems.

In the experiments, each of these classifiers were tested using all 52 extracted features with a subset of features selected using the ReliefF algorithm. The results for the three-class and four-class classification scenarios are described in Sect. 4.

4 Experimental Setup and Results

4.1 Dataset and Problem Formulation

The dataset used in this work was developed and published by the Bonn University [13]. The data set includes five subsets of signals (denoted as Z, O, N, F and S), each containing 100 single-channel EEG segments with a duration of 23.6 s. All segments were recorded using an amplifier system, digitized with a sampling rate of 173.61 Hz and 12-bit A/D resolution, and filtered using a 0.53–40 Hz (12 dB/octave) band pass filter. The normal (Basal) segments (Sets A and B) were taken from five healthy subjects. The standard surface electrode placement scheme (the international 10–20 system) was used to obtain the EEG from the healthy cases. Volunteers were relaxed in an awake state with eyes opened (A) and eyes closed (B), respectively. Both states exhibit different characteristics, the EEG readings with the eyes closed show a higher magnitude than when the eyes are opened, as well as the presence of alpha waves which are common in a relaxation state [32]. Both the Inter-Ictal and Ictal segments were obtained from five epileptic patients. The Inter-Ictal segments were recorded during seizure free intervals from the depth electrodes that were implanted into the hippocampal formations (Set C) and from the epileptogenic zone (Set D). The Ictal segments (Set E) were recorded from all sites exhibiting Ictal activity using depth electrodes and also from strip electrodes that were implanted into the lateral and basal regions of the neocortex, see Table 3.

Acharya et al. [12] suggests to formulate four distinct problems of varying degrees of difficulty, these are summarized in Table 4 as Problems 1–4. Problem 1 and Problem 2 are binary classification problems. On the other hand, Problems 3 and 4 are multi-class problems, the former considering classes A, D and E, while Problem 4 considers all five classes, the most difficult case. Note that each group (A–E) contains 100 epochs with 4097 samples each. For the first three problems, several methods have reported perfect accuracy (see the summary provided by [3].

Table 3 Summary of the Bonn data set. All classes have 100 epochs per class and 4097 samples per epoch

Data set label	Description
A	Normal, eyes open
B	Normal, eyes closed
C	Seizure free, depth electrodes, hippocampal formations
D	Seizure free, depth electrodes, epileptogenic formations
E	Epiletic activity, depth electrodes, eptic formations

Table 4 Classification problems derived from the Bonn data set

Name	Type	Classes
Problem 1	Binary	A–E
Problem 2	Binary	A, B, C, D–E
Problem 3	Multi-class	A–D–E
Problem 4	Multi-class	A–B–C–D–E
Problem 5	Multi-class	A–B–D–E

However, comparing results directly can be a bit tricky, particularly if the learning and validation process do not match, with researchers using different amounts of computation time, different data partitions, different numbers of repetitions and different evaluation criteria. In this work, we will focus on Problem 3, and leave Problem 4 (the most difficult case) for future work. However, we also include an addition 4-class problem, which we call Problem 5.

4.2 Experimental Procedure

All the feature extraction algorithms and classification models were implemented using Python 2.7 using the Scikit-learn and Numpy libraries under the Anaconda 5.0.1 distribution. The configuration of the classification models was done as follows:

- **k-NN**: *sklearn.neighbors.KNeighborsClassifier* using a neighborhood of size 10 and and the inverse of their euclidean distances as the weight function.
- **NB**: the *sklearn.naive_bayes.GaussianNB* classifier was implemented. It assumes the features follow a normal distribution.
- **Perceptron**: using the *sklearn.linear_model.Perceptron* with 500 epochs.
- **LDA**: the default *sklearn.discriminant_analysis.LinearDiscriminantAnalysis* was used.
- **QDA**: the model was constructed using the default values from the *Quadratic DiscriminantAnalysis* library.

- **MLP**: the *sklearn.neural_network.MLPClassifier* algorithm was used with 100 hidden layers and the logistic sigmoid activation function.
- **DT**: the *sklearn.tree.DecisionTreeClassifier* was initialized using the Gini impurity measure to evaluate the quality of a region split.
- **RF**: this classifier was created using the *sklearn.ensemble.RandomForest Classifier* library with 10 estimators and without limiting their depth level.
- **SVM**: Two versions of this classifier were implemented, one with the *sklearn. svm. LinearSVC* which uses a linear kernel for constructing the boundary function, and another with the *skleran.svm.SVC* class with a RBF kernel, $C = 1$ and *gamma* $= 2$ values.

As mentioned earlier two multi-class classification problems where studied in this work, Problem 3 and Problem 5, a three and five class problem respectively. In this sense, two experiments were carried out for each problem. In the first experiments, all the classifiers were trained using the 52 extracted features from the EEG samples. For the second experiment, the ReliefF algorithm was used as a feature selection stage, resulting in 15 selected features over 30 iterations of the algorithm. In order to statistically evaluate the quality of the classification models, a 10-fold cross-validation process was performed for each of the experiments.

5 Results and Discussion

Problem 3, 3 class case. The results for the first experiment, where all the features are used to train the models, are shown in Fig. 1. Note that the SVM models were not implemented in this experiment, this is because the model was not able to adjust due to the high dimensionality of the data. The RF and DT present the best performance with a median value, over the 10 fold process, of 70% the RF shows a fairly stable behavior with an interquartile range of 0.058, while the DT presents higher dispersion with an IQR of 0.0917.

For the second experiment, the ReliefF algorithm was implemented as the feature selection mechanism. The selected features for classification are listed in Table 5. The results for this experiment are presented in Table 6 and depicted in Fig. 2. The improvement is substantial, increasing the accuracy around 30% on all the classifiers, and reducing the dispersion over all the executions. The best model again is the RF, with a median value of 100%. Nevertheless, several models present some executions that achieve the same score. The second best model is QDA, with

Table 5 Subset of features identified as the most relevant by the ReliefF algorithm

Features domain		
Time	Frequency	Time-Frequency
σ_ξ	δ_μ, δ_{max}, δ_σ	RMS
nd diff	θ_μ, θ_{max}, θ_σ	REE_α
HOC 2, 3, 4	slow α_σ	REE_γ

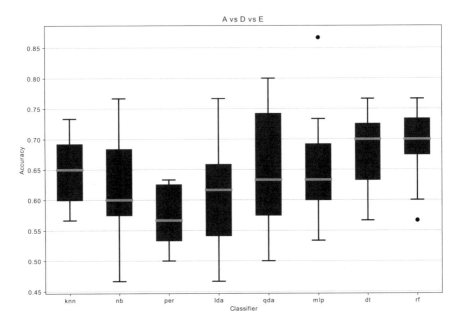

Fig. 1 Accuracy comparison of the classification models using the 52 features for Problem 3

Table 6 Accuracy comparison of the classifiers for Problem 3 using the best 15 features

Classifier	Mean	SD	Median	Minimum	Maximum	IQR
knn	0.9667	0.0394	0.9667	0.8667	1.0	0.0333
nb	0.9033	0.0567	0.9	0.7667	0.9667	0.05
per	0.9733	0.0291	0.9667	0.9	1.0	0.0333
lda	0.92	0.0521	0.9167	0.8	1.0	0.0583
qda	0.97	0.0379	0.9833	0.9	1.0	0.0333
mlp	0.9033	0.0657	0.9	0.7667	1.0	0.0833
dt	0.9433	0.0559	0.9667	0.8333	1.0	0.0833
rf	0.9633	0.0781	1.0	0.7333	1.0	0.0333
svm	0.96	0.0573	0.9667	0.8	1.0	0.0333
svm2	0.9333	0.0494	0.9333	0.8333	1.0	0.025

a median of 98.33%, this technique achieves the best mean value of 97%. These models achieve comparable results to those of the state-of-the-art presented in [3].

Problem 5, 4 class case. Similar to the previous problem, a graphical comparison of the classification performance of all models is depicted in Fig. 3. Since this problem is harder, its performance is considerably lower, all models achieve around 50% in accuracy. With four methods: NB, QDA, MLP, DT; reaching 53.75%.

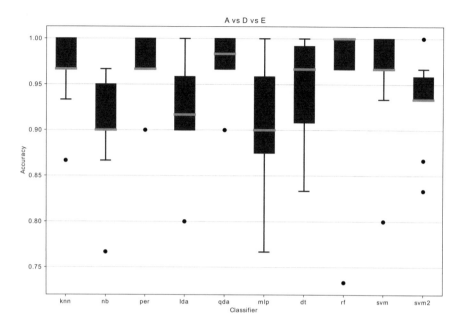

Fig. 2 Box-plot comparison of the classification models using the 15 most prominent features for Problem 3

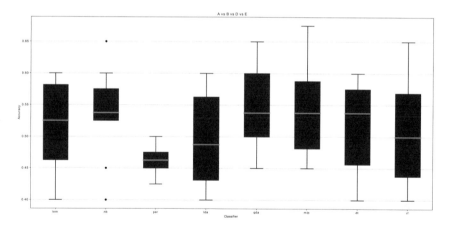

Fig. 3 Accuracy comparison of the classification models using the 52 features for Problem 5

The last experiment was done using features selected by ReliefF method (see Table 5). Similar to the other problem, the models' classifications performances is dramatically improved. All models present accuracy around 90%, as listed in Table 7 and shown in Fig. 4.

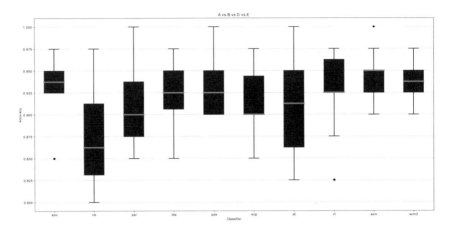

Fig. 4 Box-plot comparison of the classification models using the 15 most prominent features for Problem 5

Table 7 Accuracy comparison of the classifiers for Problem 5 using the best 15 features

Classifier	Mean	SD	Median	Minimum	Maximum	IQR
knn	0.935	0.0339	0.9375	0.85	0.975	0.025
nb	0.875	0.0548	0.8625	0.8	0.975	0.0813
per	0.91	0.0464	0.9	0.85	1.0	0.0625
lda	0.925	0.0335	0.925	0.85	0.975	0.0437
qda	0.9325	0.0336	0.925	0.9	1.0	0.05
mlp	0.91	0.0391	0.9	0.85	0.975	0.0437
dt	0.91	0.0583	0.9125	0.825	1.0	0.0875
rf	0.925	0.0447	0.925	0.825	0.975	0.0375
svm	0.9425	0.0297	0.95	0.9	1.0	0.025
svm2	0.9375	0.0202	0.9375	0.9	0.975	0.025

6   Conclusions and Future Work

This work presented an extensive study of different feature extraction and classi-fication methods in order to detect epileptic formations in EEG signals. These methods have been used for EEG analysis for the emotion classification problem, nevertheless this is the first time they are implemented for detecting epilepsy. A feature selection method (ReliefF) was used for selecting prominent features with the objective of solving two multi-class problems; a three class and a five class problem. Using 15 features, the classification models match, and in some cases outperform, the classifiers in the state of the art. After a 10-fold cross validation process, we found that the best models for the three class problems were QDA and RF, with this last one reaching perfect score as the median classification accuracy.

Similarly, the best models for the five class problems were the SVM with a linear and RBF kernels, on average reaching 94.25 and 94.25% respectively. For future work, we pretend to approach the most difficult problem for the BONN data set, the five class problem.

Acknowledgements This work was founded through the project "Clasificador de emociones 2.0 utilizando medios fisiológicos, cerebrales y conductuales" under the PEI 2017 program, with the collaboration of ITT and Neuroaplicaciones y Tecnologías S.A. de C.V.

References

1. Sotelo Arturo, Guijarro Enrique, Trujillo Leonardo, Coria Luis N, Martnez Yuliana (2013) Identification of epilepsy stages from ECoG using genetic programming classifiers. Comput Biol Med 43(11):1713–1723
2. Sotelo A (2015) Enrique D Guijarro, and Leonardo Trujillo. Seizure states identification in experimental epilepsy using gabor atom analysis. J Neurosci Methods 241:121–131
3. Flores EZ, Trujillo L, Sotelo A, Legrand P, Coria LN (2016) Regularity and matching pursuit feature extraction for the detection of epileptic seizures. J Neurosci Methods 266:107–125
4. Vézard L, Legrand P, Chavent M, Fata-Anseba F, Trujillo L (2015) Eeg classification for the detection of mental states. Appl. Soft Comput 32(C):113–131
5. Zheng W-L, Lu B-L (2015) Investigating critical frequency bands and channels for EEG-based emotion recognition with deep neural networks. IEEE Trans Auton Ment Dev 7(3):162–175
6. Jenke R, Peer A, Buss M (2014) Feature extraction and selection for emotion recognition from eeg. IEEE Trans Affect Comput 5(3):327–339
7. Fisher RS, Acevedo C, Arzimanoglou A, Bogacz A, Cross JH, Elger CE, Engel J, Forsgren L, French JA, Glynn M, Hesdorffer DC, Lee BI, Mathern GW, Moshé SL, Perucca E, Scheffer IE, Tomson T, Watanabe M, Wiebe S (2014) ILAE official report: a practical clinical definition of epilepsy. Epilepsia 55(4):475–482
8. Thurman DJ, Beghi E, Begley CE, Berg AT, Buchhalter JR, Ding D, Hesdorffer DC, Hauser WA, Kazis L, Kobau R, Kroner B, Labiner D, Liow K, Logroscino G, Medina MT, Newton CR, Parko K, Paschal A, Preux P-M, Sander JW, Selassie A, Theodore W, Tomson T, Wiebe S (2011) Standards for epidemiologic studies and surveillance of epilepsy. Epilepsia 52 Suppl 7(1):2–26
9. Eadie MJ (2012) Shortcomings in the current treatment of epilepsy. Expert Rev Neurother 12(12):1419–1427
10. Franaszczuk PJ, Bergey GK, Durka PJ, Eisenberg HM (1998) Time-frequency analysis using the matching pursuit algorithm applied to seizures originating from the mesial temporal lobe. Electroencephalogr Clin Neurophysiol 106(6):513–521
11. Kohsaka S, Mizukami S, Kohsaka M, Shiraishi H, Kobayashi K (2002) Widespread activation of the brainstem preceding the recruiting rhythm in human epilepsies. Neuroscience 115(3):697–706
12. Acharya UR, Vinitha Sree S, Swapna G, Martis RJ, Suri JS (2013) Automated EEG analysis of epilepsy: a review. Knowl Based Syst 45:147–165
13. Andrzejak RG, Lehnertz K, Mormann F, Rieke C, David P, Elger CE (2001) Indications of nonlinear deterministic and finite-dimensional structures in time series of brain electrical activity: dependence on recording region and brain state. Phys Rev 64(6 Pt 1)
14. Xie S, Krishnan S (2014) Dynamic principal component analysis with nonoverlapping moving window and its applications to epileptic EEG classification. Sci World J 1:2014

15. Kamath C (2015) Analysis of EEG dynamics in epileptic patients and healthy subjects using Hilbert transform scatter plots. OALib 02:1–14
16. Acharya UR, Molinari F, Sree SV, Chattopadhyay S, Ng KH, Suri JS (2012) Automated diagnosis of epileptic EEG using entropies. Biomed Signal Process Control 7(4):401–408
17. Ahammad N, Fathima T, Joseph P (2014) Detection of epileptic seizure event and onset using EEG. BioMed Res Int 450573. http://dx.doi.org/10.1155/2014/450573.
18. Guo L, Rivero D, Pazos A (2010) Epileptic seizure detection using multiwavelet transform based approximate entropy and artificial neural networks. J Neurosci Methods 193(1): 156–163
19. Orhan U, Hekim M, Ozer M (2011) Eeg signals classification using the k-means clustering and a multilayer perceptron neural network model. Expert Syst Appl 38(10):13475–13481
20. Tzallas AT, Tsipouras MG, Fotiadis DI (2009) Epileptic seizure detection in EEGs using time-frequency analysis. IEEE Trans Info Technol Biomed Publ IEEE Eng Med Bio Soc 13(5):703–710
21. Kovacs P, Samiee K, Gabbouj M (2014) On application of rational discrete short time fourier transform in epileptic seizure classification. ICASSP, IEEE International Conference on Acoustics, Speech and Signal Processing—Proceedings, pp 5839–5843. http://dx.doi.org/10.1109/ICASSP.2014.6854723
22. Bajaj V, Pachori RB (2012) EEG signal classification using empirical mode decomposition an d support vector machine. In: Proceedings of the International Conference on Soft Computing, pp 581–592
23. Guler N, Ubeyli E, Guler I (2005) Recurrent neural networks employing Lyapunov exponents for EEG signals classification. Expert Syst Appl 29(3):506–514
24. Durka PJ, Blinowska KJ (1995) Analysis of eeg transients by means of matching pursuit. Ann Biomed Eng 23(5):608–611
25. Durka PJ, Ircha D, Blinowska KJ (2001) Stochastic time-frequency dictionaries for matching pursuit. IEEE Trans Signal Process 49(3):507–510
26. Durka PJ, Matysiak A, Montes EM, Sosa PV, Blinowska KJ (2005) Multichannel matching pursuit and EEG inverse solutions. J Neurosci Methods 148(1):49–59
27. Hjorth Bo (1970) Eeg analysis based on time domain properties. Electroencephalogr Clin Neurophysiol 29(3):306–310
28. Hausdorff JM, Lertratanakul A, Cudkowicz ME, Peterson AL, Kaliton D, Goldberger AL (2000) Dynamic markers of altered gait rhythm in amyotrophic lateral sclerosis. J Appl Physiol 88(6):2045–2053
29. Petrantonakis PC, Hadjileontiadis LJ (2010) Emotion recognition from eeg using higher order crossings. IEEE Trans Inf Technol Biomed 14(2):186–197
30. Ackermann P, Kohlschein C, Bitsch JA, Wehrle K, Jeschke S (2016) Eeg-based automatic emotion recognition: feature extraction, selection and classification methods. In: 2016 IEEE 18th International Conference on e-health Networking, Applications and Services (Healthcom), pp 1–6
31. Fernández-Delgado M, Cernadas E, Barro S, Amorim D (2014) Do we need hundreds of classifiers to solve real world classification problems? J Mach Learn Res 15:3133–3181
32. Ball T, Kern M, Mutschler I, Aertsen A, Schulze-Bonhage A (2009) Signal quality of simultaneously recorded invasive and non-invasive EEG. NeuroImage 46(3):708–716

Big Data and Computational Intelligence: Background, Trends, Challenges, and Opportunities

Sukey Nakasima-López, Mauricio A. Sanchez and Juan R. Castro

Abstract The boom of technologies such as social media, mobile devices, internet of things, and so on, has generated enormous amounts of data that represent a tremendous challenge, since they come from different sources, different formats and are being generated in real time at an exponential speed which brings with it new necessities, opportunities, and many challenges both in the technical and analytical area. Some of the prevailing necessities lie on the development of computationally efficient algorithms that can extract value and knowledge from data and can manage the noise within in it. Computational intelligence can be seen as a key alternative to manage inaccuracies and extract value from Big Data, using fuzzy logic techniques for a better representation of the problem. And, if the concept of granular computing is also added, we will have new opportunities to decomposition of a complex data model into smaller, more defined, and meaningful granularity levels, therefore different perspectives could yield more manageable models. In this paper, two related subjects are covered, (1) the fundamentals and concepts of Big Data are described, and (2) an analysis of how computational intelligence techniques could bring benefits to this area is discussed.

1 Introduction

The emergence of the third industrial revolution was in mid-1990s, through internet technology, it brought with it a new and powerful structure that would change the world in communication and knowledge generation, with the arrival of new technologies and the transition to a new economy based on data, where it now represents a great importance, since it is considered an economic and social engine [1, 2].

Excessive data has been generated from technologies such as internet of things, social networks, mobile devices, among others. As Helbing [3] indicates, there will

S. Nakasima-López (✉) · M. A. Sanchez · J. R. Castro
Facultad de Ciencias Químicas e Ingeniería (FCQI), Universidad Autónoma
de Baja California, 22390 Tijuana, Baja California, Mexico
e-mail: sukey.nakasima@uabc.edu.mx

© Springer International Publishing AG, part of Springer Nature 2018
M. A. Sanchez et al. (eds.), *Computer Science and Engineering—Theory and Applications*, Studies in Systems, Decision and Control 143,
https://doi.org/10.1007/978-3-319-74060-7_10

be more machines connected to the internet than human users. In accordance with International Data Corporation (IDC), in 2011 were created and copied 1.8 zetta-bytes (ZB) in the world, up to 2003, 5 exabytes (EB) data were created for humanity, today that amount of information is created in two days [4].

All these technologies forming a vital part in each of the activities we carry out and it assume more important roles in our lives [5]. As is the case that Rodríguez-Mazahua et al. [6] highlights about of how Google predicted in the health public sector the propagation of flue, based searches that users did, in terms of "flu symptoms" and "treatments of flu" in a couple of weeks before there would an increase in patients arriving in a certain region with flu. This search reveals a lot about the searchers: their wants, their needs, their concerns, extraordinarily valuable information.

This exchange of information that is being generated and stored at great velocity and in exponential quantities has never before been seen throughout history, as argued by Helbing [3], the necessity emerges to mine and refine data in order to extract useful information and knowledge and to be able to make more precise forecasts, where standard computational methods can not cope with.

At present, companies and industries are aware that in order to be competitive, data analysis must become a vital factor in discovering new ideas and delivering personalized services [7]. There are a lot of potentials and high value hidden in a huge data volume, that are demanding computing innovation technologies, techniques, and methodologies to model different phenomena, with extreme precision [8].

2   Evolution of Data Analysis

With the progressive evolution of informatization, we have gone from the necessity of only storage and data management to the possibility of value extraction from data. And it is on this way where data analysis and analytics have emerged and evolved.

In 1958, Hans Peter Luhn made the first reference of business intelligence at the field of business data analysis. But it was in 1980 when Howard Dresner consol-idated the term, making references to a set of software to support the business decision making based on the collection of data and descriptive analysis, showing events that have already occurred and based on insight into the past and what is happening in the present. It is estimated that 80% of generated analytics results are descriptive, and are considered of low complexity and hindsight value [5, 9].

Subsequently the necessity for a predictive analysis was required to extract knowledge from data in the form of patterns, trends and models, and it was at the late of 1980s, when the expression of data mining emerged, whose origin is arti-ficial intelligence, that is defined as the process of discovering patterns (that must be meaningful) in data and its focus is on predictive analysis, at the same time, the expression knowledge discovery in database (KDD) also begins to be used. Soon these technique with machine learning would allow us build predictive models to determinate the outcome of an event that might occur in the future, this can lead to the identification of both risk and opportunities [5, 9, 10].

3 Emergence of Big Data

Given the arrival of ubiquitous technologies, the popularization of the world wide web and affordable personal computers, as well as other devices, we are facing a new phenomenon where the challenge is not only the storage and processing but also the analytics techniques and methodologies, that could cope to variables such as volume, velocity, and variety, the term that define these characteristics is known as Big Data.

The term was popularized in 2011 by IBM initiatives that invested in the ana- lytics market. One of the first companies to face the problem of Big Data was Google when it had the necessary to classify its pages by quality, importance, and authority, analyzing that direct clicks and that coming from other intermediate links. For this reason, Google created PageRank algorithm, but this algorithm required running in a parallel environment to cope with the challenges of volume, velocity, and variety at the same time, for that it utilized MapReduce algorithm altogether [11, 12].

On the other hand, Gandomi and Haider [11] emphasize that information that has the basic dimensions of big data are an asset for the firms that would like to create real-time intelligence from data that are changing exponentially in time. All these necessities demand cost-effective, innovative forms of information processing for enhanced insight and decision-making, that traditional data management system is not capable of handling.

There is valuable information hidden in that sea of data that could be leveraged in making-decisions, originated for clickstream from users that reveal its behavior and browsing patterns, their needs, wishes, and worries. For all these reasons, we can define that Big Data as immense data size that include heterogeneous formats (structured, unstructured and semi-structured data), that cannot be handled and analyzed by traditional databases, because the generation speed and representation limits their capacity and require advanced techniques and powerful technologies to enable the capture, storage, distribution, management, and advanced algorithms to analysis information [7, 11, 13, 14].

Another definition that was made by Hashem et al. [15] say that Big Data is a set of techniques and technologies that require new forms of integration to uncover large hidden values from large datasets that are diverse, complex, and of a massive scale.

Big Data has been described for many dimensions, where each dimension explains its complexity.

Volume: it is a major feature but not the only one, the minimum data size is considered in terabytes (equivalent to 16 millions of pictures stored in Facebook servers) and petabytes (equivalent to 1024 terabytes), the future trend is that the volume is going to increase. Transnational corporations as walmart generated batches more than one million transactions per hour, the e-commerce, sensors and social media has contributed to the generation of this huge volume of data, is estimated that Facebook has stored 260 billions of photographs, using more than 20 petabytes of storage, also Google processed a hundred of petabytes in data, as well as the electronic store Alibaba that generates dozens of transactions in terabytes per day [7, 11, 13, 16].

The existence of more data allow us to create better models since we can discover many critical variables that can help us to describe in a better way the phenomena [14, 15].

Velocity: the speed in that data is generated and processed, such that it represents a huge challenge when analyzing data in real time and having the ability to get indicators at this rate that can be useful to the decision-makers. For example, customizing the daily offer according to profile and behavior of customers, this is possible thanks to the popularity and affordability of different devices which are present in all places, like as smartphone and other mobile devices that are constantly connected to the network and provide relevant and updated information that could be leveraged to create new business strategies, such as, geospatial location, demographic information and purchasing patterns [14, 15, 17].

The tendency of improving the ability of data transmissions will continue to accelerate the velocity. In 2016, it was estimated that 6.4 billion devices were connected to the internet, using and sharing information from many sectors. Also, 5.5 billion of new devices were added in this same year, and all of them were creating and exchanging data. It is expected that for 2020 the number of devices in use and connected will be about 20.8 billion [16].

Variety: it referrers to the heterogeneous dataset that is composed of structured data (e.g. relational database and spreadsheets, among others) and these represent about 5% of all existing data, unstructured (corresponds to images, audio, and video, among others), and the semi-structured (which do not conform to a strict standard such as XML, emails, social media, and the information recovery from different devices, among others). Requiring more computing power for its efficient handling [11, 14, 16].

Garner Inc, introduced these three dimensions that are the major characterization of Big Data and are known as 3V's, however, other V's have been added that complement the complex description of this phenomena, according to Gandomi and Haider [11] and Lee [16], they identified and described the follow V's:

Value: a feature introduced by Oracle, raw data has low-value density with respect to the volume, however, we could get high value from analyzing enormous amounts of data and transforming it into strategies for the organization. The most representative benefits that obtain value are increments in revenues, reduction of operational cost, improvement to customer services, among others.

Veracity: IBM coined this characteristic that represents the distrust and latent uncertainty on the source of data generated by incompleteness, vagueness, inconsistencies, subjectivities, and latency present in data. For instance, analyzing the sentiments of people with voting age about their perception, judgment, from all kinds of comments from candidates done through social networks.

Variability: a term introduced by SAS, it refers to the variability in the rate of data streaming because of its inconsistency by intermittency, or peaks of traffic, for periods of times. Also, it denotes the complexity by connecting, combining, cleaning, and converting data collected from many sources.

4　Big Data Value Chain

All activities that taking place in an organization are known as the value chain and they have the aim to deliver a product or service to the market. The categorization of generic activities that add value to the chain allows for the organization have a better and optimized understanding about what happened in each area. All this conceptualization can be applied in virtual value chain environments such as an analytics tools that will allow us understanding of the creation of value from data technology. Doing a mapping of each phase present in the flow of information, in big data its value chain can be seen as a model of high level activities that comprise an information system [18]. The activities identified as part of value chain of Big Data are [17, 18]:

Data acquisition: It is the first activity in the chain of value, and it is referring to the collection process, filtration and cleanness of data before to placing it in whatever storage solution exists and from there do different kinds of analyses. According to Lyko et al. [19] data acquisition in the context of Big Data is governed by 4V's, where it is assumed that: a high volume, variety, velocity and a low value from the initial stage, with the aims to increase the value from collected data.

Data curation: a continuous activity of data management, because operating in each phase from process lifecycle of data, to ensure the quality and maximizing its effective use, these features are relevant and represent a strong impact on the business operations, since they influence in the process of the making-decision of an organization [20]. In this process, we seek that the data be reliable, accessible, reusable and adequate to the purpose for they were created.

Data storage: according to Chen et al. [13] in the environment of Big Data is refers to the storage and management of a huge volume of data in scalable platforms, that in turn provide reliability, fault tolerance, and availability of the data for its subsequent access. As Strohbach et al. [21] points out, the ideal characteristics of a storage system in big data will be: virtual capacity of unlimited data, high rate of random access to writing and reading, flexible and efficient, manage different models of data, support to structured and not structured data, and that can work with encrypted data for major security.

Data analysis: collected data can be interpreted in a clear form to the decision-makers, according to Yaqoob et al. [22] techniques of big data are required to make for efficient the analytics of enormous data volumes in a limited time period.

Data use: All of those tools that allow us the integration of data analysis with the business activities that are required for decision-making, to provide the organizations the opportunities to be competitive through the reduction of costs, increasing value, monitoring for many parameters that are relevant to the good functioning of the organization, generation of predictions, simulations, visualizations, explorations, and data modeling in specific domains. The sectors where data analytics have been implemented and successful are manufacturing, energy, transport, health and so on, all of them are known as industry 4.0.

5   Challenge of Big Data

Due to the complex nature of the environment of big data, many challenges are present in each stage of the data lifecycle. Also, the development of new skills, updating or the replacement of more powerful IT technologies to obtain greater performance in the process that is required. Below are listed some of them [13, 23]:

Data complexity: it is related to the characteristics that describe big data, the diversification of types, structures, sources, semantics, organization, granularity, accessibility and complicated interrelations, make it difficult to represent, understand and process the data in this context. A good representation of data allows us to obtain greater meaning, and a bad representation reduces the value of data, impeding its effective analysis. Due to the constant generation of data from different source, collection, integration and data integrity, with optimized resources both hardware and software it has become in one of the major challenges [6, 7]. To ensures quality in data, it is necessary to establish control processes, such as metrics, data evaluation, erroneous data repair strategies and so on [16].

Process complexity: related to isolating noise contained in data from errors, faults or incomplete data, to guarantee its value and utility. Reducing redundancy and compressing data to enhance its value and make them easily manageable. As Sivarajah et al. [24] points out, some of the areas that present challenges in the process are in data aggregation and integration, modeling, analysis, and interpretation.

System complexity: when it is desired to design system architectures, computing frameworks, and processing systems, it is necessary to take into account the challenge of high complexity of big data, therefore increasing requirements of processing, due to its volume, structure, and dispersed value. And that it must support large work cycles and that their analysis and delivery of results must be in real time, having as main objective the operational efficiency and the energy consumption.

6 Areas of Application of Big Data

The sectors where big data has had application and a strong positive impact, and where has overcome the storage and analysis challenges, has been:

Internet of Things: it currently represents one of the major markets. Its devices and sensors produce large amounts of data and have the potential to generate trends and investigate the impact of certain events or decisions. The development and application has been given in intelligent buildings, cyber-physical systems, as well as traffic management systems [25].

Smart grid: big data analytics allows the identification of electrical grid transformers at risk, and the detection of abnormal behavior of connected devices. It allows establishing preventive strategies with the purpose of reducing costs by correction, as well as more approximate forecasts of the demand that allow to make a better balance of the energy loads [7].

Airlines: employ hundreds of sensors on each aircraft to generate data about their entire aircraft fleet, its objective is to monitor their performance and apply preventive maintenance resulting in significant savings for the company [26].

E-health: used to customize health services, doctors can monitor the symptoms of patients in order to adjust their prescriptions. Useful for optimizing hospital administrative operations and reducing costs. One of these solutions is offered by CISCO [7].

Services: the tools of big data allow to analyze the current behavior of clients, to cross the information with historical data and their preferences, in order to offer a more effective service and to improve their marketing strategies, such as Disneyland park, who have introduced a bracelet equipped with radio frequency, which allows visitors to avoid waiting in lines and book rides, this allows to create a better experience for the visitor, attracts many more customers and increases their income [26].

Public uses: used in complex water supply systems, to monitor its flow and detect leaks in real time, illegal connections and control of valves for a more equitable supply in different parts of the city. As is the case of Dublin city council where one of the most important services is the transport and for that purpose it has equipped its buses with GPS sensors to collect geospatial data in real time and with this, through its analysis, it can optimize their routes and use of their transport, allowing fuel savings and decrease the level of pollution in the air emitted by the transport system [7, 26].

7 Computational Intelligence

Due to the increase in the complexity surrounding the data, since they are being generated in an excessive way and in very short time periods, it requires both powerful computing technology, as well as robust algorithms from which we can extract knowledge and value. In this context, a solution that can encompass the representative characteristics of the big data phenomenon is computational intelligence (CI).

According to Hill [27], CI focuses on replicating human behavior more than on the mechanisms that generate such behavior, through computation. The algorithms based on CI, allows modeling of human experience in specific domains in order to provide them with the capacity to learn and then adapt to new situations or changing environments [28–30]. Some of the behaviors it includes are; abstraction, hierarchies, aggregation of data for the construction of new conclusions, conceptual information, representation and learning from symbols. The use of CI requires a focus on problems rather than on technological development [27].

Bio-inspired algorithms are increasingly used to work and give solutions to problems with a high level of complexity, since they are intelligent algorithms that can learn and adapt as would biological organisms, these algorithms have the characteristic that they can be tolerant to incomplete, imprecise, and implicit uncertain data. They can also increase the range of potential solutions with better approximation, manageability and robustness at the same time [29–31]. Some of the technologies often associated with CI are described below:

Artificial Neural Networks (ANN): a discipline that tries to imitate the processes of learning of the brain, replicating the inaccurate interpretation of information obtained from the senses taking advantage of the benefits of the fast processing offered by computer technology. The first step toward artificial intelligence came from neurophysiologist Warren McCulloch and the mathematician Walter Pitts that in 1943 wrote a paper about how the neurons work and they established the precedent of creation of a computational model to neural network [32]. ANN are also defined as adaptive algorithms of non-linear and self-organized processing, with multiple processing units connected known as neurons, in a network with different layers. They have the ability to learn based on their inputs and adapted according to the feedback obtained from their environment [31].

Different neural network architectures have been developed, some of them are [29]:

- Hopfield network, it is the simplest of all, because it is a neuron network with a single layer. This model was proposed in 1982 by John Hopfield [33].
- Feedforward multilayer network, executes its passage forward, has an input layer, another layer for output, and in an intermediate way has the known hidden layers in which can be defined N number of them. This design was made by Broomhead and Lowe in 1988 [34].
- Self-organized networks, such as Kohonen's self-organizing feature maps and the learning vector quantizer. A paper by Kohonen was published in 1982 [34].
- Supervised and unsupervised, some networks with radial basis functions.

Its main advantages are in the handling of noise in data and good control, achieving low error rates. Neural network methods are used for classification, clustering, mining, prediction, and pattern recognition. They are broadly divided into three types, of which they are recognized; feedforward network, feedback network and self-organized networks. The characteristics of artificial neural networks are; distributed information storage, parallel processing, reasoning and self-organized learning, and have the ability to adjust non-linear data quickly. Neural networks are trained and not programmed, are easy to adapt to new problems and can infer relationships not recognized by programmers [35].

Genetic Algorithms (GA): inspired by the principles of genetics and natural selection, in order to optimize a problem, were first proposed by John Holland in 1960 [36]. Through a cost function, it tries to find the optimal value either in maximum or minimum of a given set of parameters [31]. These types of algorithms have succeeded in optimizing search systems that are difficult to quantify, such as in financial applications, industrial sector, climatology, biomedical engineering, control, game theory, electronic design, automated manufacturing, data mining, combinatorial optimization, fault diagnosis, classification, scheduling, and time series approximation [29].

In big data, GA have been applied to generate clustering, in order to have a better management of data volume, dividing data into small groups that are considered its population. Also, one of its great benefits is that they are highly parallelizable. It can be combined with K-means algorithms (created in 1957 by Stuart Lloyd [37]), the combination of GK-means will take less memory and process large volumes of data in less time and achieve very good results [38].

Fuzzy logic: all human activities have implicit uncertainty, our understanding is largely based on imprecise human reasoning, and this imprecision could be useful to humans when they must make decisions, but in turn are complex processing for computers. Fuzzy logic is a method to formalize the human capacity of imprecise reasoning. It is partial or approximate, assigning to a fuzzy set degree of truth in a set between 0 and 1 [39].

In the context of big data have the ability to handle various type of uncertainties that are present at each phase of big data processing. Also, fuzzy logic techniques with other Granular Computer techniques can be employed to the problem which can be reconstructed to a certain granular level. It would be more efficient if they are associated with other decision-making techniques, such as probability, rough sets, neural networks, among others [17]. Some of the applications of fuzzy systems have been given in; control systems, vehicle braking system, elevator control, household appliances, traffic signal control, so on [29].

Its relevance in a big data environment lies in its ability to provide a better representation of the problem through the use of linguistic variables, which facilitates the handling of volume and variety when datasets are growing exponentially and dynamically. In addition, experts can benefit from the ease of interpreting results associated with these linguistic variables [40].

Other applications are in intelligent hybrid systems, where the benefits of both fuzzy systems and neural networks are combined, enhancing the ability of the latter to discover through learning the parameters needed to process data. Being these; fuzzy sets, membership functions, and fuzzy rules. It has also been proposed to integrate to these hybrid intelligent systems, genetic algorithms to optimize parameters, adjust control points of membership functions and fine tune their fuzzy weights [41].

Granular Computing: computing paradigm for information processing, tries to imitate the way in which humans process information obtained from their environment in order to understand a problem. Can be modeled with principles of fuzzy sets, rough sets, computation with words, neural networks, interval analysis, among others. In 1979, Zadeh introduced the notion of information granulation and suggested that fuzzy set theory might find possible applications with respect to this. Its powerful tools are vital for managing and understanding the complexity of big data, allowing multiple views for data analysis, from multiple levels of granularity [42, 43].

Granules may be represented by subsets, classes, objects, clusters, and elements of a universe. These sets are constructed from their distinctions, similarities, or functionalities. It has become one of the fastest growing information paradigms in the fields of computational intelligence and human-centered systems [17].

Fuzzy logic techniques together with granular computing concepts are considered one of the best options for the process of decision-making. A fuzzy granule is defined by generalized constraints, said granules can be represented by natural language words. The fuzzification of its granules together with its values that characterize it, is the way in which human constructs its concepts, organizes, and manipulates them. They can be used to reconstruct problems with a certain level of granularity (from the finest, which could be at the level of an individual, to the coarse granules that could be at community level), the objective would be to focus on the volume, feature of big data, reducing its size and creating different perspectives that can later be analyzed and become indicators of relevance for decision-making [17, 42, 43].

Granular computing has represented an alternative solution to obtaining utility and value of big data in spite of its complexity. Because of their integration with computational intelligence theories, it can help effectively support all the operational levels that include; acquisition, extraction, cleaning, integration, modeling, analysis, interpretation, and development [42].

Machine Learning: branch of artificial intelligence, which focuses on the theory, performance, and properties of algorithms and learning systems. It is an interdisciplinary field which is related to artificial intelligence, optimization theory, information theory, statistics, cognitive science, optimal control and other disciplines of science, engineering and mathematics. Its field is divided into three subdomains, which are [44]:

- **Supervised learning**: requires training from input data and the desired output. Some of the tasks performed in data processing are classification, regression, and estimation. Of the representative algorithms in this area are: support vector machine that was proposed by Vladimir Vapnik in 1982 [45], hidden Markov model was proposed in 1966 by Baum and Petrie [46], Naives Bayes [47], bayesian network [48], among others.

- **Unsupervised learning**: only requires input data, without indicating the desired objective. The tasks of data processing that it performs are clustering and prediction, of which some of existing algorithms are: Gaussian mixture model that was created by Karl Pearson's in 1984 [49], X-means [50], among others.
- **Reinforced learning**: allows learning from the feedback received through the interaction obtained from an external environment. They are oriented to decision making and some algorithms are: Q-learning that was introduced by Zdzislaw Pawlak in 1981 [51], TD learning proposed by R. D. Sutton in 1988 [52], and Sarsa learning that was proposed by Sutton and Barton in 1998 [53].

The application of machine learning can be carried out through three phases [54]:

- **Preprocessing**: helps to prepare raw data which by its nature consists of the unstructured, incomplete, inconsistent and noisy data, through cleaning of data, extraction, transformation, and fusion of data which can be used in the learning stage as input data.
- **Learning**: uses learning algorithms to fine-tune the model parameters and generate desired data outputs from pre-processed input data.
- **Evaluation**: the performance of learning models is determined here, characteristics to which special attention is given are: performance measurement, dataset selection, error estimation, and statistical testing. This will allow you to adjust the model parameters.

The objective of machine learning is to be able to discover knowledge and to serve the decision makers to be able to generate intelligent decisions. In real life, they have applications in search engines for recommendation, recognition systems, data mining for discovery of patterns and extraction of value, autonomous control systems, among others [7]. Google uses machine learning algorithms for large volumes of disordered data.

8   Conclusions

The Big Data phenomenon confronts us with great opportunities, but also great challenges in order to achieve its benefits. Its complex characteristics invite us to rethink the ways data is managed, processed and analyzed for value, quality and relevance. It is necessary to overcome the superficial analysis that only describes a historical event and evolving to deeply analyses that allow us creating predictions and prescription that support actions and strategies to improve the decision-making in the organizations.

Traditional models and technologies cannot cope with this effectively, so it is necessary to design and develop new technologies, with greater computing power, advanced algorithms, new techniques and methodologies that serve as support to be able to have greater control of the volume, variety, and speed that are part of the complex nature of big data.

For this reason, we believe that the design and development of algorithms based on computational intelligence can be a very good way to face these challenges that characterizes Big Data by its complex nature, for it will be necessary to adapt them to the existing platforms and make many experiments, to demonstrate that they are really efficient, effective, and help us to discover new patterns, ideas and knowledge in this context as well.

References

1. Brynjolfsson E, Kahin B (2000) Understanding the digital economy: data, tools and research. Massachusetts Institute of Technology
2. Rifkin J (2011) The third industrial revolution: how lateral power is transforming energy, the economy, and the world
3. Helbing D (2015) Thinking ahead—essays on big data, digital revolution, and participatory market society
4. Akoka J, Comyn-Wattaiau I, Laoufi N (2017) Research on big data—a systematic mapping study. Comput Stand Interfaces 54:105–115
5. Thomson JR (2015) High integrity systems and safety management in hazardous industries
6. Rodríguez-Mazahua L, Rodríguez-Enríquez CA, Sánchez-Cervantes JL, Cervantes J, García-Alcaraz JL, Alor-Hernández G (2016) A general perspective of big data: applications, tools, challenges and trends. J Supercomput 72(8):3073–3113
7. Oussous A, Benjelloun FZ, Ait Lahcen A, Belfkih S (2017) Big data technologies: a survey. J King Saud Univ Comput Inf Sci
8. McKinsey & Company (2011) Big data: the next frontier for innovation, competition, and productivity. McKinsey Global Institute, p 156
9. Niño M, Illarramendi A (2015) Entendiendo el Big Data: antecedentes, origen y desarrollo posterior. DYNA NEW Technol 2(3), p [8 p]–[8]
10. Witten IH, Frank E (2005) Data mining: practical machine learning tools and techniques, vol 2
11. Gandomi A, Haider M (2015) Beyond the hype: big data concepts, methods, and analytics. Int J Inf Manage 35(2):137–144
12. Srilekha M (2015) Page rank algorithm in map reducing for big data. Int J Conceptions Comput Inf Technol 3(1):3–5
13. Chen M, Mao S, Liu Y (2014) Big data: a survey. Mobile Netw Appl 19(2):171–209
14. Kacfah Emani C, Cullot N, Nicolle C (2015) Understandable big data: a survey. Comput Sci Rev 17:70–81
15. Hashem IAT, Yaqoob I, Anuar NB, Mokhtar S, Gani A, Ullah Khan S (2015) The rise of 'big data' on cloud computing: review and open research issues. Inf Syst 47:98–115
16. Lee I (2017) Big data: dimensions, evolution, impacts, and challenges. Bus Horiz 60(3):293–303
17. Wang H, Xu Z, Pedrycz W (2017) An overview on the roles of fuzzy set techniques in big data processing: trends, challenges and opportunities. Knowl Based Syst 118:15–30
18. Curry E (2016) The big data value chain: definitions, concepts, and theoretical approaches. In: New horizons for a data-driven economy: a roadmap for usage and exploitation of big data in Europe, pp 29–37
19. Lyko K, Nitzschke M, Ngomo A-CN (2016) Big data acquisition
20. Freitas A, Curry E (2016) Big data curation
21. Strohbach M, Daubert J, Ravkin H, Lischka M (2016) Big data storage. In: New horizons for a data-driven economy, pp 119–141

22. Yaqoob I et al (2016) Big data: from beginning to future. Int J Inf Manage 36(6):1231–1247 Pergamon
23. Jin X, Wah BW, Cheng X, Wang Y (2015) Significance and challenges of big data research. Big Data Res 2(2):59–64
24. Sivarajah U, Kamal MM, Irani Z, Weerakkody V (2017) Critical analysis of big data challenges and analytical methods. J Bus Res 70:263–286
25. Ahmed E et al (2017) The role of big data analytics in internet of things. Comput Netw
26. Alharthi A, Krotov V, Bowman M (2017) Addressing barriers to big data. Bus Horiz 60 (3):285–292
27. Hill R (2010) Computational intelligence and emerging data technologies. In: Proceedings— 2nd international conference on intelligent networking and collaborative systems, INCOS 2010, pp 449–454
28. Jang J, E M, Sun CT (1997) Neuro-fuzzy and soft computing-a computational approach to learning and machine intelligence. Autom Control IEEE 42(10):1482–1484
29. Engelbrecht AP (2007) Computational intelligence: an introduction, 2nd edn
30. Kruse R, Borgelt C, Klawonn F, Moewes C, Steinbrecher M, Held P (2013) Computational intelligence. Springer, Berlin
31. Kar AK (2016) Bio inspired computing—a review of algorithms and scope of applications. Expert Syst Appl 59:20–32
32. Kumar EP, Sharma EP (2014) Artificial neural networks—a study. Int J Emerg Eng Res Technol 2(2):143–148
33. Elmetwally MM, Aal FA, Awad ML, Omran S (2008) A hopfield neural network approach for integrated transmission network expansion planning. J Appl Sci Res 4(11):1387–1394
34. Negnevitsky M (2005) Artificial intelligence: a guide to intelligent systems. In: Artificial intelligence: a guide to intelligent systems. Pearson Education, pp 87–113
35. Biryulev C, Yakymiv Y, Selemonavichus A (2010) Research of ANN usage in data mining and semantic integration. In: MEMSTECH'2010
36. Mitchell M (1995) Genetic algorithms: an overview. Complexity 1(1):31–39
37. Govind Maheswaran JJ, Jayarajan P, Johnes J (2013) K-means clustering algorithms: a comparative study
38. Jain S (2017) Mining big data using genetic algorithm. Int Res J Eng Technol 4(7):743–747
39. Ross TJ et al (2004) Fuzzy logic with engineering applications. IEEE Trans Inf Theory 58 (3):1–19
40. Fernández A, Carmona CJ, del Jesus MJ, Herrera F (2016) A view on fuzzy systems for big data: progress and opportunities. Int J Comput Intell Syst 9:69–80
41. Almejalli K, Dahal K, Hossain A (2007) GA-based learning algorithms to identify fuzzy rules for fuzzy neural networks. In: Proceedings of the 7th international conference on intelligent systems design and applications, ISDA 2007, pp 289–294
42. Pal SK, Meher SK, Skowron A (2015) Data science, big data and granular mining. Pattern Recogn Lett 67:109–112
43. Yao Y (2008) Human-inspired granular computing 2. Granular computing as human-inspired problem solving, No. 1972, pp 401–410
44. Qiu J, Wu Q, Ding G, Xu Y, Feng S (2016) A survey of machine learning for big data processing. EURASIP J Adv Sign Process 2016(1):67
45. Cortes C, Vapnik V (1995) Support-vector networks. Mach Learn 20(3):273–297
46. Baum LE, Petrie T (1966) Statistical inference for probabilistic functions of finite state Markov chains. Ann Math Stat 37(6):1554–1563
47. Rish I (2001) An empirical study of the Naïve Bayes classifier. IJCAI 2001 Work Empir Meth Artif Intell 3
48. Zarikas V, Papageorgiou E, Regner P (2015) Bayesian network construction using a fuzzy rule based approach for medical decision support. Expert Syst 32:344–369
49. Erar B (2011) Mixture model cluster analysis under different covariance structures using information complexity

50. Pelleg D, Pelleg D, Moore AW, Moore AW (2000) X-means: extending K-means with efficient estimation of the number of clusters. In: Proceedings of the seventeenth international conference on machine learning, pp 727–734
51. Pandey D, Pandey P (2010) Approximate Q-learning: an introduction. In: 2010 second international conference on machine learning and computing, pp 317–320
52. Desai S, Joshi K, Desai B (2016) Survey on reinforcement learning techniques. Int J Sci Res Publ 6(2):179–2250
53. Abramson M, Wechsler H (2001) Competitive reinforcement learning for combinatorial problems. In: Proceedings of the international joint conference on neural networks IJCNN'01, vol 4, pp 2333–2338
54. Zhou L, Pan S, Wang J, Vasilakos AV (2017) Machine learning on big data: opportunities and challenges. Neurocomputing 237:350–361

Design of a Low-Cost Test Plan for Low-Cost MEMS Accelerometers

Jesús A. García López and Leocundo Aguilar

Abstract The present work proposes a Test Plan to evaluate the performance of low-cost MEMS accelerometers, currently some of these sensors suffer from non-linearities in its outputs caused primarily by scale factors, biases and random noise, some of these factors can be compensated to a certain extent. The Test Plan is divided on three stages of testing, with each one testing different aspects of the sensor; and for those devices that are found suitable, a characterization of a basic accelerometer model is recommended to help increase their performance so that they may be able to be used in high demanding applications, such as Inertial Navigation Systems. It is important to acknowledge that this Test Plan does not ensure that every sensor will be able to be compensated, and should be primarily used for finding suitable sensors.

Keywords MEMS · Accelerometer · Calibration · Analysis

1 Introduction

In the recent trend of embedded devices there's a sensor that has achieved a ubiquitous status, this is the Micro-electro-mechanical systems (MEMS) accelerometer. Its ubiquity has been driven primarily because of the smart phone market [1]. It's inexpensive, small footprint and low power consumption. From a design point of view there is little to no drawbacks in adding it, and the information that can be obtained from the sensor can have a lot of useful applications [2–4], and can be found in a platitude of devices, ranging from toys and handheld devices, to unmanned vehicles and military weapons. They are one of the fundamental aiding sensor for determining the attitude orientation of whatever they maybe strap down to.

J. A. García López (✉) · L. Aguilar
Facultad de Ciencias Químicas e Ingeniería (FCQI), Universidad
Autónoma de Baja California, 22390 Tijuana, Baja California, Mexico
e-mail: garcia.jesus@uabc.edu.mx

© Springer International Publishing AG, part of Springer Nature 2018
M. A. Sanchez et al. (eds.), *Computer Science and Engineering—Theory and Applications*, Studies in Systems, Decision and Control 143,
https://doi.org/10.1007/978-3-319-74060-7_11

One important usage of the macro size accelerometers is in aviation applications, these sensors coupled with an angular rate sensor (gyroscope) formed the basis for an Inertial Navigation Systems (INS) [5, 6]. In principle, MEMS inertial sensors can be utilized to work as the basis for an INS; but these sensors have a few downfalls that affect their performance, mainly caused by factors such as: offset bias, cross coupling and misalignment of the sensing axes, which induce error in the measurement.

High performance inertial sensors exist and are used in current INS applications. But aligning with the vision of the father of ubiquitous computing, the future of the embedded devices should achieve the same status as a piece of paper [7], and one step toward that future is utilizing the most inexpensive parts, and not relying on military grade sensors. That's why the focus of this work will be on consumer grade MEMS accelerometers and calibration needed to enhance their performance so that they me be suitable for use in an INS.

Now, why do we need an INS? Aside from the military applications, for which a majority of the literature tends to gravitate around, location (be it zone, room level or general area) or position (a more precise point of placement, i.e. X, Y and Z) has always been a desirable for contextual information systems [5, 8, 9].

The current boom of Internet of Things (IoT) has flourish the spectrum of connected devices, and with them an out pour of data that's being collected and generated, with most of them sending it upstream which then get aggregated as Big Data [10], but most data without context is not useful. Two key contextual questions are When? and Where? The first can be easily achieved with an implementation of a Real Time Clock and a clock synchronization scheme if needed; but the latter cannot be answered as easily. All IoT devices are connected devices, with most of them being wireless, which means that most of them have a Radio Frequency (RF) Module [11].

A large body of work exist related to techniques for determining the location in an RF environment [12–15]. A major drawback of this scheme is that is requires infrastructure and prior knowledge of the distribution of the RF network is needed, limiting its ubiquity. A case could be made for a Wi-Fi or GSM networks, which have achieved a certain level of pervasiveness on most of our human environment, these are a good solution for obtaining a gross location of the device (meaning that the precise position of the device is not known, just a general area), the downside is that these Radios are power hungry; but these schemes shouldn't be undervalued, on the contrary they can be a great aid. For example, a hybrid solution of an INS and a RF Module in an RF infrastructure can be used to do Simultaneous Location and Mapping (SLAM) [16–18].

It is important to mention that the Global Positioning System (GPS) technology is the de facto solution for when the position of an object is needed [19], but it has two mayor disadvantages: (1) Since it relies on time packages being received from satellites, a clear view of the sky is needed, any obstruction will degrade its performance, which limits its use for outdoor applications only, signal is not always available, (2) They are expensive and have a huge power consumption. Which is why at the present time is it cannot be considered a true ubiquitous solution. Again, it is noteworthy to point out that GPS with an auxiliary INS have become the current trend [20], and have shown to increased their performance.

This work relates to the testing needed to be applied to a MEMS accelerometer to evaluate its performance, by characterization of the sensor and possible compensation of its measurements. It is important to note that not all low-cost sensors will be suitable, so one of the first results will be to test for their suitability for an INS application.

2 MEMS Accelerometer

Thanks to the miniaturized mechanical and electro-mechanical elements, it's been possible to recreate the structure of an inertial sensor on the micrometer scale [21]. And like its name imply, an accelerometer measures the accelerations to which it is being submitted to. A simple visualization of this sensor is to illustrate it as a proof mass that is being suspended by some spring to its reference frame. Each of this spring corresponds to an axis, so that if this spring gets compress or stretch it corresponds to a proportional output in the plus or minus scale.

The accelerometer only senses with respect to the its reference frame. We will later make use of this property to directly obtain the bias using Fig. 1d case, and later use Fig. 1c to utilize the constant of the gravity vector as a reference for our parametrization.

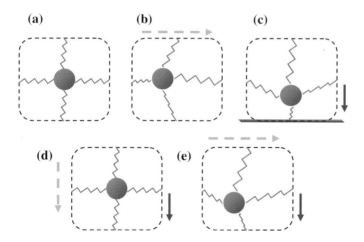

Fig. 1 **a** The proof mass suspended by the springs that are attached to the reference frame. **b** A linear acceleration is being applied. **c** At rest, only affected by gravity. **d** Free falling. **e** On Earth, it is always sensing gravity plus the linear accelerations

3   Accelerometer Model

The simplest accelerometer model can be described as:

$$\tilde{a}_i = Sa_i + b \tag{1}$$

where $\tilde{a}_i$ is the output of the accelerometer, S is the scale factor, b is the bias of the measurement, and a_i is the current acceleration being suffered by the device. But this is far too simple and does not take into account the non-linearity of the noise of the measurements and cross coupling effects. A more complete model is as follows:

$$\tilde{a}_x = (1 + S_x)a_x + B_f + n_x \tag{2}$$

where a_x is the sensitive axis, S_x is the scale factor error (usually in polynomial to include non-linear effects), B_f is the measurement bias, and n_x is the random bias. Coefficients vary with time, temperature, vibration, applied motion, power-on to power-on. The previous model is an adaptation of the model proposed by Titterton for macro size accelerometers [22], this reduced model was also used by [23].

4   Testing Stages

We propose these three stages to characterize the performance of the sensor.

1. Simple preliminary evaluation, putting the sensor in a stationary position for a short period of time, and confirming that the output is compatible with the manufacturer's specifications.
2. Static testing needed for obtaining the compensation parameters of the sensor from a multi-position test.
3. Dynamic testing, sensor is subjected to a few controlled motions.

The novel part of the design of this test plan is to use simple and low-cost test equipment, keeping with the trend of using low-cost sensors. During the testing it is important to maintain an accurate log, recording the details of the test and their result.

4.1   Data Logging

The basic setup is going to be:

1. A microcontroller capable of sampling the sensor at its highest rate.
2. The MEMS Sensor(s).

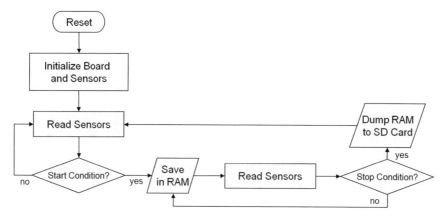

Fig. 2 Fast acquisition logic

3. Non-Volatile Memory with sufficient memory to log the data for at least 4 h. A 4 GB SD Card should be ample enough.
4. Temperature Sensor, for parametric profiling of temperature change.
5. Small portable power supply, e.g. Li-Po battery with a regulator, USB power bank.
6. And optionally an RF Module can be used to record the data directly into the PC in the majority of the tests.

The basic dynamic is going to be to just read the sensors at a leisure pace and either log it on the on-board non-volatile memory or, if the RF Module is present, send it to a PC so that it can be stored by a receiving application.

But there is going to be a real time constrain on the Free-Falling Test, that will later be described, were the sensors have to be sampled at the fastest rate possible for a very short amount of time. To be able to meet that deadline the following process shown on Fig. 2 is advised.

4.2 Preliminary Evaluation

As mention before, this test is the observation of the sensor on stationary position, it doesn't matter its orientation, what we want to ascertain is that it is functioning within the specifications. This test should record the output of the sensor for a 15-min period immediately after the Power-On state. What should be noted on this stage is the sensitivity is inside specs and the warm-up trends. If the output is not within the specifications or if the output does not stabilize after a 5 min warm up, that that sensor is not usable and should not be considered for further testing.

4.3 Static Testing

4.3.1 Multi-position Test

This test involves putting the sensor in multiple specific positions, and using the constant gravity vector for our testing. Typically, an expensive test equipment such as a Dividing Head is needed. We propose the utilization of a polyhedron with 18 faces (octadecahedron), which is basically just a tumbling sphere as shown on Fig. 3.

The sensor is attached to one of the faces of the solid. This polyhedron is generated by first making a sphere, and then making cuts with a depth of 10% of its diameter, at every 45°. The size of the polyhedron will depend on the size of the sensor board, which needs to me attached to one of the faces of the mounting device. Once the sensor is mounted we can start the recording, rotating the polyhedron on each of its faces and leaving it stationary for a short time (1 min should suffice), capturing the behavior of each axis when submitted to 0, −1, −0.5, 0.5 and 1 g. Should be noted that a sensor could mounted on every face, so it would possible to evaluate 18 sensors at the same time, if necessary. Finally, the mounting device should be place on a rubber mat to minimize the vibrations from the room or bench.

The mounting device could theoretically be made of any material, but if a 3D printer is available, then that might be the most effective way. An example of Fig. 3 can be found in the reference section, which can be downloaded and printed; but the dimension of the sensor board should be taken into account and scaled the given solid as needed. If no 3D Printer is available, then the cited solid can be used as a reference model.

The geometry of the solid needs to be rectified, any imperfections of the mounting device will only produce an incorrect parametrization. Any obvious defect protruding from be print should be shaved or sanded, every edge and face of the polyhedron should be flat. And every face should be perpendicular to the face

Fig. 3 Multi-position mounting device [29]

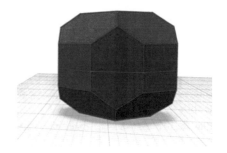

on the opposite side; to insure this property, a Vernier caliper can be utilized to confirm that every corner of the face is of equal distance to its opposite face, rectifying that the faces are truly perpendicular. The difference should within 1 mm, if it is out of this range, then it means that the 3D Printer Z-axis tolerance are too high and cannot be used to print the mounting device.

With the captured data, the following parameters can be found:

1. Scale Factor.
2. Scale Factor linearity.
3. Null bias error (though this will be corroborated on the following stage).

In order to obtain the previous parameters, we must first adjust our orientation since the mounting device might not be truly parallel to the ground, this may be caused by any offset gradient of the surface on which the device is sitting (bench, desk, floor, etc.), meaning that surface will not be truly perpendicular to the Earth's gravity vector and this will skew our measurements.

Thankfully the sensor has been static and the only acceleration that is observed is the gravity vector, we can use this vector to calculate our inclination and adjust our readings. Figure 3 illustrates an acceleration vector on the accelerometer, it's angles and the relation with the sensing axes.

The computation of the Rotation Matrix only needs to be done once, on the starting position of the device (Fig. 4).

From the previous illustration we can express the rotation angles between the axes and the resulting vector as:

$$\cos(A_{xr}) = \frac{R_x}{R}; \quad \cos(A_{yr}) = \frac{R_y}{R}; \quad \cos(A_{zr}) = \frac{R_z}{R} \tag{3}$$

Fig. 4 Angles of the resulting vector of an accelerometer's intertial reference frame

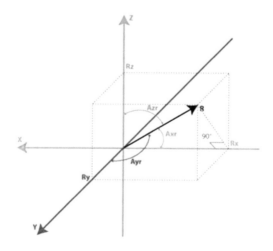

Which can also be expressed as:

$$A_{xr} = \cos^{-1}\left(\frac{R_x}{R}\right); \quad A_{yr} = \cos^{-1}\left(\frac{R_y}{R}\right); \quad A_{zr} = \cos^{-1}\left(\frac{R_z}{R}\right); \quad A_{zr} = \cos^{-1}\left(\frac{R_z}{R}\right) \tag{4}$$

With these angles we can populate our Direction Cosine Matrix to represent the orientation of the device [24].

$$\theta = A_{xr}, \quad \phi = A_{yr}, \quad \psi = A_{zr}$$

$$R = \begin{bmatrix} \cos\theta\cos\psi & \sin\phi\sin\theta\cos\psi - \cos\phi\sin\psi & \cos\phi\sin\theta\cos\psi + \sin\phi\sin\psi \\ \cos\theta\sin\psi & \sin\phi\sin\theta\sin\psi\cos\phi\cos\psi & \cos\phi\sin\theta\sin\psi - \sin\phi\cos\psi \\ -\sin\theta & \sin\phi\cos\theta & \cos\phi\cos\theta \end{bmatrix} \tag{5}$$

And finally, we can adjust our recorded measurements, multiplying our measurement vector by our Rotation matrix:

$$\hat{a} = R\tilde{a} \tag{6}$$

where, $\bar{a}$ is our adjusted output that has one axis parallel to the gravity vector. Now there should only be one axis being affected at every 90° and should read a 1 and -1 g for its maximum and minimum value respectively. Any discrepancy is mainly case by the scaling factor and should be noted to be later compensated.

A simple approximation can be done by using the following formula:

$$B_i = \frac{\bar{a}_i^{up} + \bar{a}_i^{down}}{2} \tag{7}$$

In which i is the corresponding sampled axis, $\bar{a}_i^{up}$ is the average of $\hat{a}_i$ when the axis is pointing up, $\bar{a}_i^{down}$ is the average of $\hat{a}_i$ when the axis is pointing down, b_i is the corresponding bias parameter. Ideally $\bar{a}_i^{up}$ and $\bar{a}_i^{down}$ should be opposites and the resulting sum should be cero, any discrepancy is caused by the bias.

And for the scale factor, something similar can be done:

$$S_i = \frac{\bar{a}_i^{up} - \bar{a}_i^{down} - 2g}{2g} \tag{8}$$

where here again $\bar{a}_i^{up}$ and $\bar{a}_i^{down}$ should be opposites and the resulting difference should be equal to 2 g, any difference will get assigned as the scale factor S_i.

An example of a single axis of the Multi-Position Test rotation sequence can be appreciated on Fig. 5. Only the intervals of data that are marked by a roman

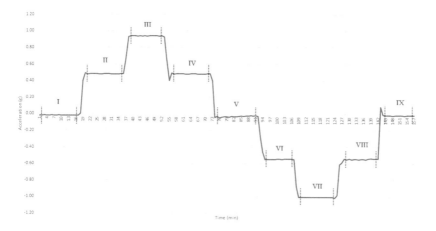

Fig. 5 A sample of the rotation sequence, with annotated intervals of the useful data

numeral are meant to be used in our parametrization, that means that any data that is logged while the device is in transition to the next rotation is to be ignored; the device needs to be static.

The rotation sequence is generated by rotating only over a sensing axis that is perpendicular to the ground. A full rotation is achieved by stopping at every 45° increments, which are the faces of the multi-position device. Each of this rotation sequence is described as follows:

I. Initial position of the device, though it is of no importance to start on this particular position, the only aspect that matters is to start and end on the same position, completing a full 360° rotation. We will mark this position as the 0° orientation of the mounting device, and looking at the example shown on Fig. 5 we can notice that the sensor axis is perpendicular to the gravity vector.

II. 45° position. The output of the sensor should be close to 0.5 g.

III. 90° position. The sensor axis is pointing up, and a 1 g measurement should be sensed. Averaging this interval will give us $\bar{a}_i^{up}$.

IV. 135° position. Again, the sensor should output 0.5 g.

V. 180° position. The sensor axis is again perpendicular to the ground, output should be 0 g.

VI. 225° position. The sensor should output −0.5 g.

VII. 270° position. The sensor axis is pointing down, and a −1 g measurement should be sensed. Averaging this interval will give us $\bar{a}_i^{down}$.

VIII. 225° position. Again, the sensor should output −0.5 g.

IX. 360° position. We have returned to the starting position. At this point, we have only captured the rotation sequences of two of the axes, since the third axis was always orthogonal to the gravity vector. To capture the last axis, we first need to rotate 90° in a perpendicular direction (left or right) of the completed rotation sequence. Now we can repeat from step I through VII and finalize our multi-position test capture.

The linearity of the scale factor can be observed by plotting only the average of the previous eight intervals (because the last one is a repeated position). If a non-linearity in the captured data is observed then that means that the scale factor that was found in Eq. 8 is not the best approximation. A Least Square method can be used to fit this parameter over the captured data, as shown on [23].

To achieve this, we must first rewrite our accelerometer output equation in a matrix form as:

$$
\begin{bmatrix} \tilde{a}_x \\ \tilde{a}_y \\ \tilde{a}_z \end{bmatrix} = \begin{bmatrix} m_{xx} & m_{xy} & m_{xz} \\ m_{yx} & m_{yy} & m_{yz} \\ m_{zx} & m_{zy} & m_{zz} \end{bmatrix} \begin{bmatrix} a_x \\ a_y \\ a_z \end{bmatrix} + \begin{bmatrix} b_x \\ b_y \\ b_z \end{bmatrix} \tag{9}
$$

Which can be rewritten as:

$$
\begin{bmatrix} \tilde{a}_x \\ \tilde{a}_y \\ \tilde{a}_z \end{bmatrix} = \underbrace{\begin{bmatrix} m_{xx} & m_{xy} & m_{xz} & b_x \\ m_{yx} & m_{yy} & m_{yz} & b_y \\ m_{zx} & m_{zy} & m_{zz} & b_z \end{bmatrix}}_{M} \underbrace{\begin{bmatrix} a_x \\ a_y \\ a_z \\ 1 \end{bmatrix}}_{a} \tag{10}
$$

where m are the misalignment factors between different axis, and the scale factors between the same axis (the elements that are on the diagonal). The other elements are the same that in Eq. 2.

From the previous rotation sequence, we now the following six basic positions and their corresponding theoretical values, we will describe them as:

$$
a_1' = \begin{bmatrix} 0 \\ 0 \\ 1g \end{bmatrix} \quad a_3' = \begin{bmatrix} 0 \\ 1g \\ 0 \end{bmatrix} \quad a_5' = \begin{bmatrix} 1g \\ 0 \\ 0 \end{bmatrix}
$$
$$
a_2' = \begin{bmatrix} 0 \\ 0 \\ -1g \end{bmatrix} \quad a_4' = \begin{bmatrix} 0 \\ -1g \\ 0 \end{bmatrix} \quad a_6' = \begin{bmatrix} -1g \\ 0 \\ 0 \end{bmatrix} \tag{11}
$$

With this we can now form the adjustment matrix:

$$
A = \begin{bmatrix} a_1' & a_2' & a_3' & a_4' & a_5' & a_6' \\ 1 & 1 & 1 & 1 & 1 & 1 \end{bmatrix} \tag{12}
$$

And finally, the reading of the sensors:

$$
u_1 = \begin{bmatrix} \tilde{a}_x \\ \tilde{a}_y \\ \tilde{a}_z \end{bmatrix}_{Z-\text{axis up}} \quad u_3 = \begin{bmatrix} \tilde{a}_x \\ \tilde{a}_y \\ \tilde{a}_z \end{bmatrix}_{Y-\text{axis up}} \quad u_5 = \begin{bmatrix} \tilde{a}_x \\ \tilde{a}_y \\ \tilde{a}_z \end{bmatrix}_{X-\text{axis up}}
$$
$$
u_2 = \begin{bmatrix} \tilde{a}_x \\ \tilde{a}_y \\ \tilde{a}_z \end{bmatrix}_{Z-\text{axis down}} \quad u_4 = \begin{bmatrix} \tilde{a}_x \\ \tilde{a}_y \\ \tilde{a}_z \end{bmatrix}_{Y-\text{axis down}} \quad u_6 = \begin{bmatrix} \tilde{a}_x \\ \tilde{a}_y \\ \tilde{a}_z \end{bmatrix}_{X-\text{axis down}} \tag{13}
$$

With which we form the sensor output matrix:

$$
U = \begin{bmatrix} u_1 & u_2 & u_3 & u_4 & u_5 & u_6 \end{bmatrix} \tag{14}
$$

Now we can lastly solve for M of Eq. 10, using the renowned least square method:

$$
M = UA^T \left(AA^T \right)^{-1} \tag{15}
$$

4.3.2 Long Term Stability Test

For this test, the accelerometer output is logged in a static state for a prolong period of time, for at least 5 h, but can be left for days or even weeks for a thorough stability profile. Since the interest is in the trend of the measurements over a wide time spectrum, there is no need for a fast data output rate, and can be left at a 1 or 2 Hz sampling rate. And should be repeated with each axis parallel to the gravity vector. It is crucial to logged the temperature to which the sensor is being expose to, since there is a direct correlation between the output data and the temperature, because of the capacitive sensing technology of the low-cost MEMS sensor.

4.3.3 Repeatability Test

Finally, the last Static Test evaluates the repeatability of the measurements. For this we need to repeat the following test for at least 12 times:

1. Turn off the device, rotate it to some other face of the polyhedron y leave it static for a while.
2. Turn on the device and start recording the data. Repeat step 1.

The sensor should show repeatability of its measurements on each power-on event when place on the same orientation as previously logged. If not, then it cannot be used for our purpose. Ideally this test should be carried out on different days.

4.4 Dynamic Testing

4.4.1 Free Fall Test

As illustrated on Fig. 1, when the accelerometer is on free fall, no acceleration should be sensed by the accelerometer, because the proof mass is falling at the same rate as the reference frame, and so the mass should be perfectly centered on the reference frame, and it should be outputting 0. So, substituting $\tilde{a}_x = 0$ in Eq. (2) leaves us with the following:

$$\tilde{a}_x = B_f + n_x \qquad (16)$$

So, any measurement while on free fall is just the bias plus noise. This test should be done on the highest sampling rate possible, so that we can extract the trend of the data and filter out the noise. The resulting trend is bias, and it should be almost the same as the bias that was obtain on the multi-position test; if there is a discrepancy, it may because by an erroneous scale factor and it should be corrected with this newly observe bias. Meaning that:

$$B_i = \bar{a}_i^{fall} \qquad (17)$$

In which i is the corresponding sampled axis, $\bar{a}_i^{fall}$ is the average of $\tilde{a}_i$ when the device was on free-fall; and B_i is the resulting bias. Again, this bias should corroborate what was found on Eq. (7).

To carry out this test an aerodynamic case (Fig. 6) was used to minimize the drag force while free-falling. The shape of the case is modeled with an ogive tip to

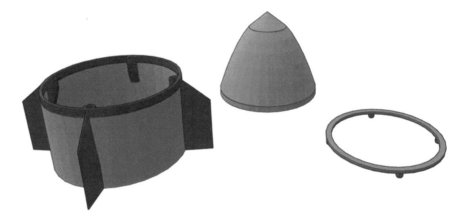

Fig. 6 Free fall mount [30]

reduce drag, and rear fins to eliminate any rotation while failing. Furthermore, a counter weight must be placed on tip of the case to remove any wobble. All of this is done to achieve a stale fall and that no other acceleration is sensed.

Three to five meters fall at the maximum sampling rate should suffice (see diagram of Fig. 2).

It is vital that the fall be dampened, for example, by a receiving extra padded mat or a stretchy cloth or netting. Otherwise the receiving shock on impact could damage the sensor permanently. This is later discussed on the Conclusion section.

4.4.2 Centrifuge Test

This test should be done with caution since it could damage the sensor. The test will subject the sensor to higher values of acceleration and evaluate the linearity of the scale-factor that was previously found in the static tests. And to continue with the trend of low cost equipment, for this test we need to build a centrifuge that is basically a Brushless DC motor and an Electric Speed Controller (ESC) and a Microcontroller to be used as the ESC master (Fig. 7).

Careful logging will be needed to record the speed, and consecutively the resulting force being applied to the sensor by means of:

$$F = M\omega^2 r \tag{18}$$

where F is the resulting force, M is the mass of the object, ω the angular velocity and r is the radius of the centrifuge. Extra precaution will be needed no to over expose the sensor to a higher force that it is rated. This could permanently damage the sensor.

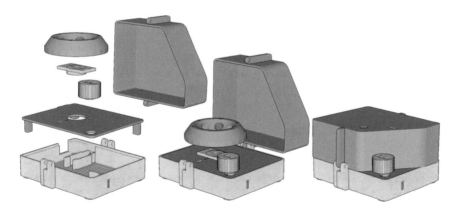

Fig. 7 Example of a low-cost centrifuge [31]

5 Conclusions

The proposed test plan is to evaluate the performance of a low-cost MEMS sensor and ascertain their suitability for a higher requirement application, e.g. an Inertial Navigation System. While at the same time parameterize a basic model of the accelerometer to help improve its performance so that it can be used in more demanding applications.

In practice, the test can be run recursively, using the estimated error from one test to update the parameters and optimize the model of the accelerometer. Instead of Least-Square methods for fitting these parameters, Kalman filtering techniques are also a prefer method for this type of sensors [25, 26].

The noise parameter was never profiled on this test plan since it is a stochastic phenomenon and it is out of the scope of this present plan, but is the subject of the future work. Some simple schemes would be to use heuristics, thresholds, and filtering, like a Savitzky-Golay filters [27], with a high sampling rate to minimize the impact of the random noise in our system.

Future work will consist of submitting a wide variety of low-cost MEMS accelerometer to this test plan, and categorizing them based on their suitability and performance.

Another factor that will worsen the performance of the sensor is any strong shock the sensor received, be it by a fall or an impact by another object. This shock can damage the internal mechanical structure of the senor, changing the behavior of the sensor, or even rendered it useless.

On a final note, one key aspect that should be taken into account is the aging of the sensor. The MEMS accelerometer functions in a vacuum, this vacuum is provided by the packaging of the sensor in which it is encased into [28]. As time goes on, the seal of the packaging will degrade, and with this the performance of the accelerometer will also worsen, and render the compensation of the parameters ineffective. To account for this, the Test Plan could be repeated and evaluate if the sensor is still suitable; if it is, then to update its compensation parameters.

Acknowledgements We would like to thank the MyDCI program of UABC, our friends, colleges and instructors; and for the financial support provided by our sponsor CONACYT, with the contract grant number: 383573.

References

1. Perlmutter M, Robin L (2012) High-performance, low cost inertial MEMS: a market in motion! In: Proceedings of the 2012 IEEEION position location navigation symposium (PLANS), pp 225–229
2. Foerster F, Fahrenberg J (2000) Motion pattern and posture: correctly assessed by calibrated accelerometers. Behav Res Methods 32:450–457

3. Sukkarieh S, Nebot EM, Durrant-Whyte HF (1999) A high integrity IMU/GPS navigation loop for autonomous land vehicle applications. IEEE Trans Robot Autom 15:572–578. https://doi.org/10.1109/70.768189

4. Zhou S, Shan Q, Fei F, et al (2009) Gesture recognition for interactive controllers using MEMS motion sensors. In: 4th IEEE International conference on nanomicro engineered and molecular systems (NEMS 2009), pp 935–940

5. Beauregard S, Haas H (2006) Pedestrian dead reckoning: a basis for personal positioning. In: Proceeding 3rd workshop positioning, navigation and communication, pp 27–35

6. Kaur A, Balsundar P, Kumar V, et al (2016) MEMS based inertial navigation system: an exploratory analysis. In: 2016 5th International conference wireless networks and embedded systems (WECON), pp 1–6

7. Weiser M (2002) The computer for the 21st century. IEEE Pervasive Comput 99:19–25. https://doi.org/10.1109/MPRV.2002.993141

8. Blasch E, Herrero JG, Snidaro L et al (2013) Overview of contextual tracking approaches in information fusion. In: SPIE defense, security, and sensing. International society for optics and photonics, p 87470B

9. Caron F, Duflos E, Pomorski D, Vanheeghe P (2006) GPS/IMU data fusion using multisensor Kalman filtering: introduction of contextual aspects. Inf Fusion 7:221–230

10. Khan R, Khan SU, Zaheer R, Khan S (2012) Future internet: the internet of things architecture, possible applications and key challenges. In: 2012 10th International conference frontiers of information technology, pp 257–260

11. Sun Q, Liu J, Li S et al (2010) Internet of things: summarize on concepts, architecture and key technology problem. J Beijing Univ Posts Telecommun 3:1–9

12. Bouzera N, Oussalah M, Mezhoud N, Khireddine A (2017) Fuzzy extended Kalman filter for dynamic mobile localization in urban area using wireless network. Appl Soft Comput 57:452–467. https://doi.org/10.1016/j.asoc.2017.04.007

13. Liu Y, Dashti M, Zhang J (2013) Indoor localization on mobile phone platforms using embedded inertial sensors. In: 2013 10th Workshop positioning navigation and communication (WPNC), pp 1–5

14. Yang Q, Pan SJ, Zheng VW (2008) Estimating location using Wi-Fi. IEEE Intell Syst 23:8–13. https://doi.org/10.1109/MIS.2008.4

15. Zhao F, Luo H, Geng H, Sun Q (2014) An RSSI gradient-based AP localization algorithm. Commun China 11:100–108. https://doi.org/10.1109/CC.2014.6821742

16. Huang J, Millman D, Quigley M, et al (2011) Efficient, generalized indoor wifi graphslam. In: IEEE international conference on robotics and automation (ICRA 2011). IEEE, pp 1038–1043

17. Li Y (2016) Integration of MEMS sensors, WiFi, and magnetic features for indoor pedestrian navigation with consumer portable devices. University of Calgary

18. Zhuang Y, Syed Z, Li Y, El-Sheimy N (2016) Evaluation of two WiFi positioning systems based on autonomous crowdsourcing of handheld devices for indoor navigation. IEEE Trans Mob Comput 15:1982–1995

19. Hofmann-Wellenhof B, Lichtenegger H, Collins J (2001) Global positioning system: theory and practice. Springer-Verlag, Wein

20. Jiang C, Zhang S-B, Zhang Q-Z (2017) Adaptive estimation of multiple fading factors for GPS/INS integrated navigation systems. Sensors 17:1254. https://doi.org/10.3390/s17061254

21. Boser BE, Howe RT (1996) Surface micromachined accelerometers. IEEE J Solid-State Circuits 31:366–375

22. Titterton D, Weston JL (2004) Strapdown inertial navigation technology. IET

23. Park M, Gao Y (2008) Error and performance analysis of MEMS-based inertial sensors with a low-cost GPS receiver. Sensors 8:2240–2261. https://doi.org/10.3390/s8042240

24. Premerlani W, Bizard P (2009) Direction cosine matrix IMU: theory. Diy Drone, USA, pp 13–15

25. Beravs T, Podobnik J, Munih M (2012) Three-axial accelerometer calibration using Kalman filter covariance matrix for online estimation of optimal sensor orientation. IEEE Trans Instrum Meas 61:2501–2511. https://doi.org/10.1109/TIM.2012.2187360

26. Hellmers H, Norrdine A, Blankenbach J, Eichhorn A (2013) An IMU/magnetometer-based Indoor positioning system using Kalman filtering. In: 2013 International conference on indoor positioning and indoor navigation (IPIN), pp 1–9
27. Savitzky A, Golay MJ (1964) Smoothing and differentiation of data by simplified least squares procedures. Anal Chem 36:1627–1639
28. Bao M (2005) Analysis and design principles of MEMS devices. Elsevier, New York
29. 3D design tumbling Sphere. In: Tinkercad. https://www.tinkercad.com/things/4unglYdH7gi. Accessed 16 Oct 2017
30. 3D design Snap-on-Rocket. In: Tinkercad. https://www.tinkercad.com/things/3PLpsYUz5VS. Accessed 16 Oct 2017
31. Thingiverse.com F.Lab's DIYbio Centrifuge by F_Lab_TH. https://www.thingiverse.com/thing:1175393. Accessed 16 Oct 2017

Evaluation of Scheduling Algorithms for 5G Mobile Systems

Christian F. Müller, Guillermo Galaviz, Ángel G. Andrade, Irina Kaiser and Wolfgang Fengler

Abstract One of the key elements of the fifth generation (5G) of mobile communication systems is the support of a large number of users communicating through a wide range of devices and applications. These conditions give rise to heterogeneous traffic offered to the network. In order to carry such traffic in a wireless network, the design and development of schedulers capable of considering the conditions of each user is needed. In this chapter a Model Based Design (MBD) and Model Based Testing (MBT) approach are used to implement and evaluate different scheduling algorithms that consider the Quality of Service (QoS) requirements of each user as well as the individual channel conditions. The development process is achieved through a hardware platform consisting of an FPGA and a System on Chip in order to provide an emulation environment. The development process as well as the results obtained (in terms of throughput and fairness) through the evaluation of Maximum Rate (MR), Round Robin (RR), Proportional Fair (PF) and a proposed novel UE-based Maximum Rate (UEMR) scheduling algorithms are presented. Using MBD together with MBT the developed scheduling algorithms can be further enhanced and exported to real world applications.

1 Introduction

During the last few decades, mobile communications have contributed to the economic and social development of most countries. In addition, we have seen a proliferation of smartphones and new mobile devices that support a wide range of

C. F. Müller
Programmable Solutions Group, Intel Deutschland GmbH, Feldkirchen, Germany

G. Galaviz · Á.G. Andrade (✉)
Facultad de Ingeniería, Universidad Autónoma de Baja California, Mexicali, Baja California, Mexico
e-mail: aandrade@uabc.edu.mx

I. Kaiser · W. Fengler
Department of Computer Science and Automation, Computer Architecture and Embedded Systems Group, University of Technology Ilmenau, Ilmenau, Germany

© Springer International Publishing AG, part of Springer Nature 2018
M. A. Sanchez et al. (eds.), *Computer Science and Engineering—Theory and Applications*, Studies in Systems, Decision and Control 143,
https://doi.org/10.1007/978-3-319-74060-7_12

applications and services ranging from messages and image transfer to video streaming for entertainment or video calls. Unlike previous generations of cellular networks, the fifth generation (5G) mobile communication system is envisioned to support a multitude of devices [e.g., Internet of Things (IoT)], and applications more extensive and enriched through richer content being delivered in real-time (e.g. high definition video streaming (4 K), augmented reality (AR), high definition gaming, and 3D contents) [1–3]. According to the International Telecommunication Union (ITU), with these emerging applications demanding higher data rates, and ultra-reliable and low latency communications, 5G networks are expected to have the peak data rate of around 10 Gbps and latency around 1 mseg, which is a 100-fold improvement over current fourth generation (4G) networks. Therefore, data traffic over heterogeneous 5G networks will not be evenly distributed; it will be extremely high in super dense areas (such as shopping malls, and stadiums), and there will be variations in the traffic volume depending on the time, location, application, and type of device. Due to this, the 5G mobile network operators will be forced to improve the performance of their networks, design them with heterogeneous traffic conditions in mind, and look for mobile technologies with higher spectral efficiency [4].

The Third Generation Partnership Project (3GPP) announced in October 2015 the plan to further evolve Long Term Evolution (LTE) to prepare the path towards 5G through LTE-Advanced Pro (LTE-A Pro) marketed as "4.5G" [5–7]. The major advances achieved with the completion of LTE-Advanced Pro (Release 13 and beyond) include Machine-Type-Communications (MTC) enhancements, public safety features, carrier aggregation enhancements, unlicensed spectrum usage, licensed shared access, single cell-point to multi-point and work on latency reduction [8].

For increasing spectral efficiency, reducing the latency and handling several classes of traffic and types of devices with drastically varying Quality of Service (QoS) requirements, LTE uses scheduled-based channel access instead of the contention-based scheme as used by Wi-Fi systems. However, designing radio resource allocation or scheduling techniques to increase the current data rates over LTE-A Pro is one of the challenging problems [9, 10]. Parameters such as number of users, channel conditions, type of traffic class and so on play an important role while designing a scheduling algorithm. Various algorithms on scheduling and resource allocation have been proposed in the literature [11–13], which can be classified as channel unaware, channel aware/QoS unaware, channel aware/QoS aware or energy aware [14].

The main goal reported in this document is the development and evaluation of schedulers using a Model Based Design (MBD) approach based on a real-time evaluation system emulating the downlink communication between a LTE-Advanced (LTE-A) network cell (eNB) and several User Equipments (UEs). An FPGA-based real-time evaluation system for Maximum Rate (MR), Round Robin (RR), Proportional Fair (PF) and a proposed novel UE-based Maximum Rate (UEMR) algorithms is developed. A realistic traffic model is developed to generate heterogeneous traffic to be used as input for the *radio resource scheduler*. The measured results are presented in real-time to the tester. Furthermore, it allows running for a specific time generating *evaluation* data for network throughput, UE-based fairness and the fairness of the different User Groups (UG). The objective is to design and simulate the best effective resource allocation strategy for LTE-based technology by comparing different algorithms and hence designing the best-case scheme for resource scheduling.

1.1   Long Term Evolution (LTE)

The radio transmissions in LTE are based on the Orthogonal Frequency Division Multiplexing (OFDM) scheme. In particular, the Single Carrier Frequency Division Multiple Access (SC-FDMA) is used in uplink transmissions and the OFDM Access (OFDMA) is used in downlink transmissions [15, 16]. LTE frames are 10 ms in duration. Each frame is divided into 10 sub-frames, where each sub-frame's duration is 1 ms called Transmission Time Interval (TTI). Sub-frames are further divided into two slots, where Tslot = 0.5 ms. In LTE, resource allocation for users is implemented in both the time domain and the frequency domain. In the time domain, resources are allocated every TTI, i.e. every 1 ms. In the frequency domain, LTE supports a high degree of bandwidth flexibility allowing for an overall transmission bandwidth ranging from 1.4 MHz up to 20 MHz. Packet scheduling (PS) in LTE is the process of allocating Physical Resource Blocks (PRBs), to users/devices over the shared data channel. PS is performed by the base stations (eNB). The overall goal of PS is to make the most efficient use of radio resources in order to maximize the cell capacity without neglecting specific QoS and fairness requirements [17].

On the other hand, LTE-Advanced (LTE-A) has been created based on the definition of 4G systems. LTE-A started a new era of improved network capacity as well as higher bandwidth availability to each user. LTE-A specifications support operation with bandwidths of up to 100 MHz, taking advantage of multi-carrier functionalities such as Carrier Aggregation (CA) [18, 28], and also other techniques such as advanced Multiple-Input-Multiple-Output (MIMO) [19, 20], heterogeneous networks (HetNet) [21], and improved interference mitigation techniques including interference avoidance and coordinated multi-point (CoMP) transmission schemes [22–25].

LTE and LTE-A are still being deployed in many parts of the world. In fact, these technologies have shown a huge enhancement of data rates and in spectral efficiency. However, the increasing demand of mobile data traffic, leads to the development of new wireless technology releases. LTE-A Pro refers to LTE enhanced with the new features included in Release 13 (which was finalized in March 2016), Release 14 (which was released in January 2017 and expected to be finalized before September 2017), and onwards [9]. One of the salient features of LTE-A Pro is extending LTE into the 5 GHz unlicensed spectrum. For this, Licensed Assisted Access (LAA) is proposed as a variant of LTE in unlicensed bands [26]. LAA uses licensed spectrum as the primary carrier for signaling (control channels) and to deliver data for users with high Quality-of-Service (QoS) requirements. CA, as a key enabler to aggregate multiple spectrum opportunities of unlicensed spectrum, is used to add secondary component carriers (CC) in the unlicensed spectrum to deliver data to users with best-effort QoS requirements [27, 28]. For this, 3GPP introduced in LTE-A Pro an extension of the CA framework beyond five component carriers (CCs) and enable massive carrier aggregation to transmit on much larger aggregated bandwidth, improving user experience through increased bandwidth. In HetNets, this CA functionality helps to maintain the basic connectivity and mobility under the macro cell coverage while small cells called "Add-on" cells achieve higher throughput performance and larger capacity. Another of the key technical topics in the LTE-A Pro specification releases is the uplink latency reduction. This is for improving throughput of Transmission Control Protocol (TCP) traffic in the downlink. In this case, a packet scheduler in base stations is a required function [29].

1.2 Scheduling Fundamentals

Radio Resource Management (RRM) lies at the heart of wireless communication networks, since it aims at guaranteeing the required Quality of Service (QoS) at the user level [30, 31]. The objective of the RRM is to utilize the limited radio-frequency spectrum resources and radio network infrastructure as efficiently as possible [9, 17, 32]. In general, RRM is divided into two categories: scheduling and resource allocation [20, 33]. The scheduler normally decides which user must be served and determines the number of packets that should be scheduled in the current time. The resource allocator decides which resource block (RB) is assigned to the selected user, and determines the number of RBs requires to satisfy the user's QoS requirements. The scheduler is the main element to the overall systems performance, even more in networks with high traffic volume. A scheduler is in charge of scheduling user data transfers in the time domain, all through the same wide band radio channel. Furthermore, not only the network load and the radio channel condition have to be considered, but also guaranteeing the QoS requirements of the used applications [15]. LTE offers nine different QoS classes [34].

The PS decides on the order between two or more packets, which will be transmitted [35, 36]. For this task, the traffic in the network is classified into traffic classes also called QoS classes. The different network services require different quality of service requirements e.g. emails and standard web applications having lower requirements as voice and video calls [37–39]. Each user is allocated a number of so-called RBs in the time frequency grid. User's bit rate depends upon the number of RBs and modulation scheme used. The number of RBs and the kind of resource blocks a user gets allocated is purely dependent on the scheduling algorithm. The basic operation of the scheduler is to determine when each user will make use of the shared spectrum resource. The decisions made by the scheduler to determine when the user can use the channel are based on different factors. These factors include (but are not limited to) channel conditions, type of service requested, channel availability, overall network throughput and fairness.

PS can be divided into two types: channel-independent scheduling and channel-dependent scheduling (dynamic scheduling) [40–42]. Channel independent scheduling does not take channel conditions into consideration. On the other hand, channel-dependent scheduling, in general, can achieve better performance by allocating resources based on channel conditions with optimal algorithms. Therefore, when channel-state information is available in channel-dependent PS, better performance can be achieved by adapting to varying channel conditions in wireless networks when compared to PS with unavailable channel-state information. Channel-dependent (dynamic) PS uses the Channel Quality Indicators (CQI) reports from active users/devices for scheduling PRBs. These reports are used by the eNB for scheduling decisions in the downlink and the uplink directions every TTI. The scheduler can use one of 15 different CQI values for mapping between the CQI and the modulation scheme to be used for transmission. The scheduling decisions are not only based on CQI reports, but also on many parameters such as the amount of data queued in the buffers of users/devices reported via Buffer Status Reports (BSRs), Hybrid Automatic Repeat Request (HARQ) and QoS attributes that are defined for each application. The scheduling decision also depends on the Link Adaptation (LA) measurements of the instantaneous Signal-to-Noise Ratio (SINR), which is used for selecting the Modulation and Coding Scheme (MCS) for data transmission [43].

2 Model-Based Design Approach

The use of embedded systems is found in all fields of the human life from standard home appliances to machineries and complex control systems. In the development of wireless embedded systems and the corresponding mobile wireless communication technologies, testing is a main task that ensures their functionality and correctness. Nowadays, about 50% of all costs are spent for testing purposes [44]. In the classic development strategies, hardware and software are tested in very late stages of product development. At the same time detected errors in late stages of the development are very costly and time consuming to correct. For that reason *Model-Based Design* (MBD) [45] and the *Model-Based Tests* (MBT) [46] have been established. The MBD tries to simplify the complex system development, while MBT allows to test in very early stages of system development.

MBD is a mathematical and visual method of addressing problems associated with designing complex control, signal processing, and communication systems. The MBD paradigm is significantly different from traditional design methodology. Rather than using complex structures and extensive software code, designers can use MBD to define plant models with advanced functional characteristics using continuous-time and discrete-time building blocks. Not only is the testing and verification process enhanced, but also, in some cases, hardware-in-the-loop simulation can be used with the new design paradigm to perform testing of dynamic effects on the system more quickly and much more efficiently than with traditional design methodology.

The goal of MBT is to minimize the risk of failures during the integration stage, when hardware and software are combined [47]. Models or simulations to ensure the dependencies between them replace components and other system elements. Another important point is the test automation. The models to be tested, have to be explicit and formal unique. Only then complex systems can be used for test automation. In addition, the use of models shall improve the transparency, uniqueness and comprehensibility of the test specifications.

3 System Development

Modern wireless communication systems like LTE-Advanced or LTE-A Pro make use of complex signal processing algorithms to optimize spectrum efficiency and increase data rates as required by the standardization bodies. Field Programmable Gate Arrays (FPGAs) are devices that provide a platform with the advantages of robustness and speed of hardware, but without the need for time-consuming hardware development processes. One strategy that manufacturers have followed is to provide with intellectual property models that implement different signal processing functions and algorithms into their FPGAs. If models are available for different tasks, MBD is an approach that reduces the time needed to develop and test algorithms for wireless communication systems given that the implementation can be tested in early development stages.

Moreover, given the speed advantages of hardware-implemented algorithms over computer simulations, models can be used to build evaluation platforms. This can be in the form of co-simulation models or full hardware implementations for system evaluation. In both cases, MBD and MBT together form the strategy to follow in order to reduce the time required for development and testing. Particularly, Software Defined Radio (SDR) platforms are widely used for wireless communication system design and testing [48, 49]. Furthermore, the research for the optimal MBD approach for SDR is a currently discussed topic [50].

The real-time evaluation system used in this work is a Small Form Factor Software Defined Radio Development Platform (SFF SDR DP) from Lyrtech. The algorithms for mobility management and scheduling within the Radio Resource Management (RRM) subsystem will be implemented in the hardware platform. With this real-time evaluation system, it is possible to compare different scheduling algorithms with each other not only on performance metrics such as throughput, fairness and delay, but also on complexity and power consumption.

The SFF SDR DP from Lyrtech is designed to develop applications for SDR and to evaluate algorithms or software applications. The board contains three modules: the digital processing module, the data conversion module and an RF module. The digital processing module contains a Field Programmable Gate Array (FPGA) and a System on Chip (SoC) device. The data conversion module includes dual-channel analog-to-digital and digital-to-analog converters. The RF module allows transmitting and receiving various frequency bands [51]. The software tools needed for system development using the SFF SDR platform are Mathworks Matlab, FPGA Development Suite, Code Composer Studio and the Model Based Design Kit from Lyrtech.

3.1 The Radio Resource Management Model

The RRM contains several tasks such as *power control*, *cell search*, *cell reselection*, handover of UEs, Radio Link Monitoring and *Radio Admission Control* (RAC), connection monitoring, connection establishment and if necessary re-establishment, location services, Self-Optimizing Networks (SONs), Inter-Cell Interference Control (ICIC) and *scheduling* [16, 52].

The RAC is managing if a new bearer with a Guaranteed Bit Rate (GBR) can be established or not concerning the current network load situation. Therefore, the access can be denied in a heavy load situation. The RAC observes the current radio load situation using the radio link monitoring [53]. In addition, the **Channel Quality Indicator** (CQI) manager receives the CQI reports of the active UEs in the cell. The CQI reports the current channel condition between the UE and the eNB. These reports are mainly used for scheduling decision. The scheduler performs the scheduling algorithm taking into account the current channel condition shared by all users. Therefore, the scheduler decides at which UE the next packet is sent using the current CQI report of all UEs in the cell. A general rule might be a UE with a better CQI is given priority over one with a low channel condition. The accurate decision is based on the current QoS requirements of a UE and their CQI. The scheduler selects a resource block (RB) for the next scheduled element. Usually more than one UE and packet are scheduled at the same time given the radio link structure. The scheduler performs a decision every TTI = 1 ms. For evaluation, the well-known radio-scheduling algorithms Round Robin and Maximum Rate (max-C/I) are implemented. Furthermore, the Multicarrier Proportional Fair and the Satisfaction-Oriented Resource Allocation (SORA) will also be implemented for comparison [54]. Another important algorithm closely related with the scheduler and the RRM is the **Hybrid ARQ manager** (HARQ). The HARQ handles the packets, which has to be retransmitted. Within one TTI the scheduler has to decide in cooperation with the HARQ if a new packet is transmitted or if another packet is repeated (see Fig. 1). These implementations will be performed using MDB and MBT. For each implementation, several iterations stages were needed. Each iteration stage involves addition of functions or improvements to the models.

The **first iteration model** of the RRM involves two steps. In the first step, a simple First-In First-Out (FIFO) block was used as a scheduler. The FIFO simply forwards the packages without making any scheduling decisions. This allowed

Fig. 1 Radio resource
scheduler and HARQ
manager

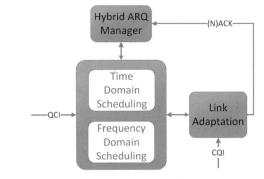

Fig. 2 Implementation
diagram of the first iteration
for RRM model

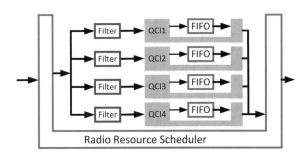

testing the first iteration of the traffic model with the RRM model. At this step, an
environment model is not needed. After successful testing, the second step of the
model involved adding a classifier or filter for the different traffic classes as shown
in Fig. 2. The generated traffic is split into four different queues. Inside the queues a
FIFO algorithm is performed. The different queues are allowed to send a packet
using the round robin algorithm. The round robin handles every queue with the
same priority and therefore with the highest possible fairness among queues. At this
point, the RRM model does not use any information about radio channel condition.

The **second iteration model** considers the environment model. The environment
model provides information used to separate the traffic into different Channel
Quality Indicator (CQI) levels. Four different CQI levels for each available radio
channel are considered. Nevertheless, the traffic is separated into five different paths
for every available RB. This means that each CQI path has its own traffic. Since
there are only four different CQI levels implemented in the environment model and
five available RBs, the CQI2 (level 2) makes use of RB2 and RB3. Afterwards
within the CQI paths the traffic is filtered in the four different QCIs. The traffic class
scheduler works like the one in the first iteration model. The model is fair regarding
the CQIs and the QoS Class Indicator (QCI).

The **final iteration** adds the functionality for reordering priorities, so it is now
possible to select for each packet the best radio link based on the current channel
radio condition for each Resource Block (RB). Furthermore, timing and sampling

issues are addressed. This final iteration step is not a final scheduling algorithm. Nevertheless, it contains all elements needed to implement the three evaluated scheduling algorithms.

3.2 Environment Model

The **environment model** generates inputs for traffic model as well as for the RRM model. It creates information regarding the UEs in the LTE-A network. The traffic model needs information about how many UEs is in the simulated network and the QoS class to which each UE belongs. Furthermore, the RRM model needs information about the QCI for each UE in the network to be able to schedule the traffic accordingly. The RRM algorithms are the second main part in the hardware development together with the traffic model. The radio resource scheduler is the main part of the RRM; it is also referred as channel-dependent scheduling. The scheduler controls the shared resources among the UEs and thus among the users. It allocates the shared resource for one or more UEs at each TTI. The scheduling algorithms use the radio channel conditions of each UE and channel to share the available spectrum as effectively as possible. Therefore, sometimes a scheduler with a high fairness level, meaning each UE is getting closely the same amount of resources, is called a highly effective scheduler.

The Round Robin scheduler is an example for a fair scheduler. While other scheduler focus on a max throughput by the network and called therefore effective at maximizing throughput. The Max C/I is a scheduling algorithm that focuses on maximizing the network throughput. However, for evaluating a radio resource scheduler it is necessary to verify if the QoS requirements of the requested services are fulfilled. In the following section, the implemented algorithms are explained in more detail.

4 Experimental Setup

4.1 Maximum Rate Scheduler

The *max C/I* algorithm shown in Fig. 3 is often called *Maximum Rate* (MR) [55]. The MR takes the current radio channel condition or radio link quality instantaneously into account and schedules the user with the currently best radio channel condition, which is for every UE within the cell different. However, there will be always a UE having almost the perfect channel conditions. The scheduler aims to maximize the throughput of the network and serves the UE with the currently best channel condition. The algorithm can be mathematically expressed as follows;

Fig. 3 Implementation
diagram for Max C/I
scheduler

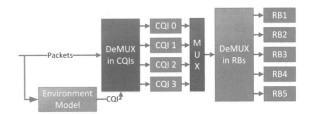

$$k = arg\ max_i\ R_i,$$

where R_i is the current data rate for UE i. This algorithm is not fair in all situations. For instance, if a UE is closer to the eNB it will have higher probability of having a better radio channel condition than a UE further away. If it is the case and the radio channel condition for one UE is for a longer time in poor conditions it might never be scheduled and therefore the MR leads into "*starvation*" for any UE with poor channel condition. This scheduling algorithm is not optimal in the terms of QoS and fairness. Nevertheless, it is the best in terms of network throughput optimization.

The MR is implemented to compare the performance of other algorithms against the maximum network throughput achieved by the max C/I. In Fig. 3, it can be seen the packets are arriving at the RRM model. Every packet belongs to a specific user. Therefore, the RRM model reads the current radio channel condition from the environment model. That traffic model contains one value for each UE for all RBs. That ensures that the network throughput for the entire network is maximized. Otherwise, the implementation would only create the maximum rate for each RB. After the current radio channel condition is known the packets are classified regarding the current CQI. The Multiplexer (MUX) block puts the packets back in one queue. However, the reading algorithm first takes all packets from the CQI0 queue before transmitting packets out of the CQI1 queue. The same occurs with CQI2 and CQI3. In the end, the packets are equally shared over the five RBs. In each RB queue a traffic classification is implemented. Nevertheless, it works in this scheduling algorithm with a Round Robin scheduler. Therefore, no prioritization is performed, but it allows for easy modification of the MR scheduler for further scheduler development or test case creation.

4.2   Round Robin Scheduler

The implemented **Round Robin (RR) scheduler** is shown in Fig. 4. This scheduler does not consider the current radio channel condition to decide which UE is served next. It shares the resources equally in the matter of time with the all UEs in the current cell. Nevertheless, the RR is not fair in terms of offering the same QoS to all UEs. Achieving that would mean to prioritize the UEs with low radio channel conditions by serving them more often as UEs with better radio channel conditions

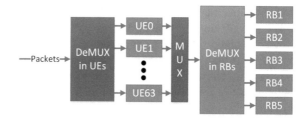

Fig. 4 Implementation diagram of Round Robin scheduler

to insure the same level of QoS. Because it serves all UEs in the cell regardless of their channel condition, the overall throughput is lower in direct comparison to the MR. The reason for that the less bits can be used for transporting data at the same time. At low radio channel conditions more information to protect the data (channel coding) has to be added and further more less bits are transported at the same using a low modulation scheme.

The Round Robin is a typical example in computer and IP-based networks. It is always used for giving a maximum level of fairness. As shown in Fig. 4, fairness focuses on the UEs served in the cell. The *RB implementation* therefore splits the traffic into the maximum number of UEs within the cell each having its own FIFO queue. A counter activates each queue once per period. Afterwards the selected packets are placed equally in the five Resource Blocks (RBs), because the RR scheduler does not use the current radio channel condition of the UEs for scheduling decisions. In the last step, each RB queue separates the traffic into the different traffic classes and schedules the classes equally. This implementation shows that the RR is a very theoretic scheduling algorithm, because of its high amount of individual queues for each UE in the current cell.

4.3 *Proportional Fair Scheduler*

The implementation diagram of Proportional Fair (PF) scheduler is shown in Fig. 5. The PF is referred as either single-carrier proportional fair [56–58] or multicarrier proportional fair [59] depending on the environment used. The LTE-Advanced network is a multicarrier network; therefore the described PF algorithm corresponds

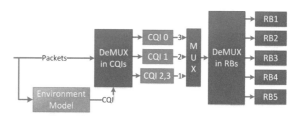

Fig. 5 Implementation diagram of Proportional fair scheduler

to the multicarrier variant. This scheduling algorithm is far more complex than the other two previously explained. The goal of the PF is to take the current radio channel condition into account and with that improving the overall network throughput, but at the same time ensuring the same average or minimum throughput. Those features will avoid starvation of UEs with low radio channel conditions from the MR scheduler as well as improve the total network throughput in comparison to the RR algorithm. The PF scheduler assigns the shared spectral resources to the UE with the relative best radio channel conditions. That can be mathematically expressed that at each TTI; the UE k is selected corresponding to;

$$k = arg\,max_i \frac{R_i}{\underline{R_i}}$$

where R_i is the current data rate for UE i and R_i is the average data rate of UE i. The average data rate is calculated over a certain period. That period needs to be sufficiently short to avoid being influenced by the long-term difference of radio channel conditions while at the same time longer as the time for the quality variations for short-term variations. The PF is a scheduling algorithm that tries to maximize the total network throughput while supporting a minimum data rate for all UEs. Therefore, the algorithm serves the UEs with better channel conditions more often than those with poor radio channel conditions. The implementation of the widely described algorithm demands a radio channel model focusing on the bitrate. For its implementation none radio channel model or adaptive code rate and modulation scheme was developed. Therefore, the functionality of the PF algorithm is different without losing the focus of a constant minimum bit rate and maximizing the transport of packets for better UEs with better channel conditions.

In Fig. 5, it can be seen that the packets arrive at the RRM model and the current channel condition of all RBs for the UE are compared against each other. The packet is marked with the RB number of the best channel condition. Therefore, the packet will later be transmitted over that RB. If the current packet has the CQI0 then the packet is placed in the first queue. That queue is allowed to send three packets out in one period. A period contains six packets. The CQI1 queue sends two packets. In the case that the radio channel condition is poorer than CQI1 it is placed in the last queue, which sends only one packet. That guarantees a minimum packet rate even for UEs with poor conditions. The packets in CQI2, 3 are still marked with the belonging number of the RB, while the packets in CQI0 and CQI1 are transmitted over the marked RB. That implementation sends more packets over good radio channel conditions while as well ensure a minimum packet rate. Inside the RB block, packet queues are separated in the four traffic classes (QCI). The different queues are sending packets using a Round Robin algorithm.

Fig. 6 Implementation
diagram of UE-based
Maximum Rate (UEMR)
scheduler

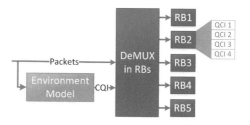

4.4 UE-Based Maximum Rate Scheduler

Additionally to the schedulers described in the literature, in this chapter a proposed novel algorithm was integrated in this RRM evaluation system. This scheduler is called UE-based Maximum Rate (UEMR), and it is shown in Fig. 6. This scheduler works similar to the PF scheduler, however in this algorithm the packets are always sent over the link with the best conditions. The best condition case is when the system is in the lowest value of CQI3, then it will transmit over an available RB. Inside the RB queues, the traffic is separated into the four traffic classes. The different traffic classes have diverse priorities of packet transmission. The QCI1 sends four packets while the QCI4 sends only one packet per period. A period contains ten packets. Therefore, the scheduling algorithm maximizes the throughput for each UE-based on the QCI value of the different RBs. Furthermore, it is the only implemented algorithm using the traffic class for deciding which packet will be sent first. This algorithm is an evolution of the classical algorithms.

5 Results Analysis

The first analysis performed is the throughput of the entire system using the different schedulers shown in Fig. 7. Furthermore, the throughput of each RB is compared with each algorithm. The entire evaluation was performed using three different simulation cycles. The simulation cycles are 1024, 10,240 and 102,400. The results of the simulation showed that the system needs a specific time until it is working with realistic behavior. The reason for that is that the queues of the RRM scheduler have to be filled. Otherwise, it is possible that a queue is allowed to send packets while being empty. Furthermore, the traffic model used in this work needs to create a specific number of packets. Therefore, the simulation time should be at least around 10,000 cycles; otherwise the effects described will occur.

The **Maximum Rate (MR)** algorithm serves the most packets in the evaluation system. That is why this algorithm is very simple and because of that very fast to execute. Therefore, the time a packet spends within the scheduler is lower as in others. The MR algorithm achieves at 1024 simulation cycles only a percentage of 92.47% of the total throughput. The remaining 7% is either stored in the internal

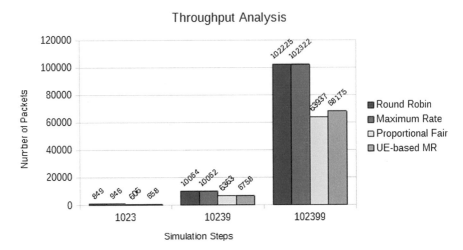

Fig. 7 Total network throughput in terms of number o packets

memories or discarded because a queue was already completely filled. The assessment block does not analyze the number of discarded packets. However, a short calculation for this example can be made. The MR contains four FIFO memories being able to save max 16 packets each. That means 64 packets can be stored within the queues and the scheduler serves 956 out of 1023 created packets. Furthermore, less than ten packets are within the scheduler buffers. Therefore, it can be assumed that the scheduling algorithm discarded around 60 packets. The percentage of served packets in comparison to the created packets is increasing to 98.27% at 10,239 simulation cycles and to 99.93% at 102,390 simulation cycles. That means, the MR achieves almost a perfect throughput.

The **Round Robin (RR)** scheduler is a more complex algorithm because the traffic is split into a queue for each UE. In the beginning, the RR performs at a lower rate as compared to the MR. The RR serves only 82.99% at 1024 simulation cycles. While at 10,240 already 98.29% are served and at 102,400 simulations cycles more than 99.8% are achieved. This result of RR shows the importance of filling the queues at the beginning for the scheduling algorithms. In terms of network throughput MR and RR algorithms show a very good performance result in the tested cases.

The **Proportional Fair (PF)** and the **UE-based Maximum Rate (UEMR)** do not achieve a lower throughput performance. The main reason for that is that the queues are weighted. The PF sends more packets to the UEs with the best and good channel conditions as to the other UEs. The Proportional Fair (PF) and the UE-based Maximum Rate (UEMR) are not achieving a lower throughput performance mainly because their queues are weighted. The PF sends more packets to the UEs with the best and good channel conditions as to the other UEs. The algorithm is not adaptive, meaning empty queues are not skipped in the algorithm selecting the next queue sending a packet. The UEMR algorithm weights the QCI with higher demands, therefore more QCI0 packets are sent. Nevertheless, the algorithm does not skip an empty queue. That is why, the network throughput is in the measured case lower than the RR or MR. The UEMR has a slightly increased packet rate in each simulation. At 102,400 simulation steps, the UEMR achieves a percentage throughput of 66.58%, while the PF comes to 64.44%. The reason for the higher throughput of the UEMR is that the implementation is less complex. Therefore, the general delay of sending a packet is lower.

Furthermore, the **throughput for each RB** achieved by each algorithm is compared. It is expected that the RR shares the resource equally between the different RB. In Fig. 8, the different throughputs per each RB are illustrated. The Round Robin fulfills the expectations and shares the RB equally. Furthermore, the MR distributes the packets equally to all RBs. The reason for that is the simple environment model. The radio channel conditions are the same for all RBs. Therefore, the packets shared independently over the RBs.

The PF and the UEMR algorithms using comparator algorithms to choose the best radio channel condition at the moment of packet arriving. That comparator is optimized to make a fast decision, therefore if the radio channel condition for RB1 has the best possible value the packet will always be sent over that link even if another RB offers the same radio channel condition. Therefore, more packets are transmitted over the first RB and the second RB. Furthermore, it is possible that the radio channel conditions in the other RB are better when the comparison takes place.

Another analysis is regarding the fairness of the RRM schedulers per UE. In a real-life network, each UE in the network is requesting certain amount of traffic in form of packets. In this system, each UE requests the same amount of packets the only exception is the first UE in each group related to the LFSR design. In Fig. 9, the scheduled packets for each UE are presented.

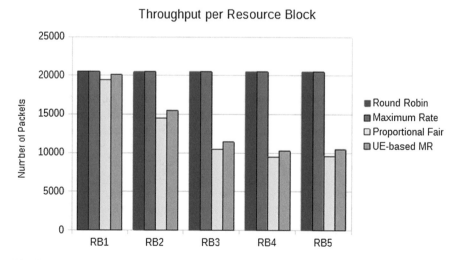

Fig. 8 Total network throughput per RB

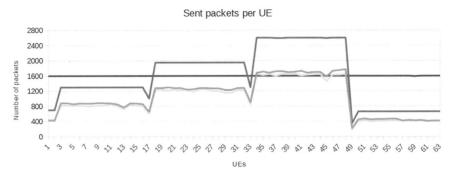

Fig. 9 Total of send packets per UE

The **RR algorithm** is working as forecasted; every UE receives the same amount of packets. Therefore, the algorithm is working as planned and is not taking network specific information like traffic classes or radio channel conditions into account for scheduling decisions. The **MR algorithm** achieved the highest throughput. In the case of fairness in serving each UE equally, the MR is not fair. In Fig. 9, the MR sends the incoming packets almost the same way out. The lower transmitted packets are the first UE of each UG related to the LFSR design. The four different amounts of packets are linked to the UG generated by the traffic model. In the tested system, the MR algorithm performs fair distribution for each UE within a UG. Nevertheless, in networks with a higher traffic load or even in congestion situation, the MR will not be fair.

The **PF** and the **UEMR** algorithm perform similarly as both are using the complex environment model for scheduling decisions. Therefore, each UE within one UG is not receiving the same amount of data. The difference between the diverse UEs and their radio channel conditions has influence on the serving rate. Once again, the effect of not skipping empty queues has a significant influence for the throughput of each UE. In addition, specific UE seem to be treated less fair as others. Furthermore, the first UE of each UG gets fewer packets as the other UE within the UG.

Another analysis of the algorithm is the **fairness of serving the UGs**. Each UG has a specific amount of requested traffic. The scheduling algorithms are not optimized for selecting the UGs for scheduling decisions. The fairness between the different UG can be interpreted differently. In this case, a scheduling algorithm is working fair, if the UGs are served equally towards their requested packets. Therefore, in Fig. 10 the four different RRM scheduling algorithms are compared with the created packets per UG in percentages. The RR scheduler treats each UE equally and therefore each UG is served with the same amount of packets.

Because of such condition the RR is not fair at serving the UG in dependency of their requested packets. In the tested case, the MR scheduler achieves a high level of fairness in serving the different UGs. The deviation, in all four cases, ranges 0–0.03%. The standard deviation of the different UGs for the PF scheduler

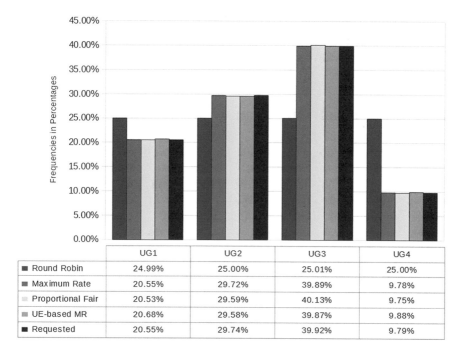

	UG1	UG2	UG3	UG4
■ Round Robin	24.99%	25.00%	25.01%	25.00%
■ Maximum Rate	20.55%	29.72%	39.89%	9.78%
▨ Proportional Fair	20.53%	29.59%	40.13%	9.75%
▨ UE-based MR	20.68%	29.58%	39.87%	9.88%
■ Requested	20.55%	29.74%	39.92%	9.79%

Fig. 10 Fairness evaluation of serving the UGs

is 0.09%. The UEMR comes to standard deviation of 0.04% between the different UGs compared with the requested packets. The PF and the UEMR are serving the UGS within the network almost in the amount as the packets were requested. The MR as well receives a very high fairness. Nevertheless, in cases of a higher loaded or even in a congestion situation the MR will not be able to serve the UG in this way.

The presented results show the functionality of the evaluation system. The four different algorithms are behaving as expected; nevertheless, the analysis in this document covers only a small number of possibilities. In the following, a few ideas for evaluating RRM scheduling algorithms that can be performed are presented. For this evaluation, the traffic model as the traffic generator works at the same clock speed as the RRM model. Therefore, a congestion situation will not happen. Furthermore, the network is not heavily loaded and the scheduling algorithms achieve very similar results. If the traffic generation clock is increased, more packets will be created and behavior of the different algorithms will vary depending on the chosen situation. In addition, a congestion situation can be achieved, where it is not possible to schedule the arriving packets anymore. The comparison of the different algorithms under those situations will create valuable results.

Furthermore, specific network situations such as high use of voice or video calls can be tested. The effect of a heavy loaded real-time video traffic within the network can be analyzed, situation possible during a large sport event like the football world cup or the Olympic Games. The usage of the evaluation platform allows many different scenarios to be analyzed and can be used to develop new RRM algorithms as well as to assess and improve existing ones.

6 Conclusion and Future Trends

In this chapter, an evaluation system for assessment and evaluation of RRM scheduling algorithms was presented. The growth of traffic in 5G mobile and wireless communication systems makes it necessary to optimize the development of new algorithms for using the available spectrum more efficiently. A real-time evaluation platform for RRM algorithms was designed, developed and tested. The developed system allows the user running complex analysis of RRM scheduling algorithms. The current state of the evaluation system can be modularly extended adding more functionality. The system proposed needs to be extended for a radio channel model or adaptive coding and modulation schemes. The functionality as a FPGA-based co-simulator allows faster simulation results against software simulations. From a technical perspective it seems to be ultimate challenging to provide uniform service experience to users under the premises of future small-cell scenarios or network densification. However, heterogeneous networks and small cell deployments will bring significant challenges in resource allocation and scheduling for 5G systems. Research on optimizing resource allocation for users with heterogeneous data rate requirements and services will be needed. The system developed here is a powerful evaluation platform offering a test bed for the implementation of various scheduling algorithms. The algorithms can easily be imported in the system design and evaluated using the evaluation systems assessment block. Also, there are a wide variety of opportunities for future research works in emerging 5G networks, such as; introduction of mm-wave spectrum, antenna array design, massive MIMO, low latency and QoE, energy efficiency, and novel multiplexing schemes.

References

1. Agiwal M, Roy A, Saxena N (2016) Next generation 5g wireless networks: a comprehensive survey. IEEE Commun Surv Tutorials 18(3):1617–1655
2. Panwar N, Sharma S, Sing A (2016) A survey on 5G: the next generation of mobile communication. Phys Commun 18:64–84
3. Shafi M et al (2017) 5G: a tutorial overview of standards, trials, challenges, deployment, and practice. IEEE J Sel Areas Commun 35(6):1201–1221
4. Mumtaz S, Morgado A, Huq KMS, Rodriguez J (2017) A survey of 5G technologies: regulatory, standardization and industrial perspectives. Digital Commun Netw. https://doi.org/10.1016/j.dcan.2017.09.010

5. Haidine A, Hassani SE (2016) LTE-a pro (4.5G) as pre-phase for 5G deployment: closing the gap between technical requirements and network performance. Paper presented at International Conference on Advanced Communication Systems and Information Security (ACOSIS), Marrakesh, Morocco, 17–19 October 2016
6. Rost P et al (2016) Mobile network architecture evolution toward 5G. IEEE Commun Mag 54(5):84–91
7. Akyildiz I, Nie S, Chun LS, Chandrasekaran M (2016) 5G roadmap: 10 key enabling technologies. Comput Netw 106:17–48
8. Schwarz S, Rupp M (2016) Society in motion: challenges for LTE and beyond mobile communications. IEEE Commun Mag 54(5):76–83
9. Suganya S, Maheshwari S, Latha YS, Ramesh C (2016) Resource scheduling algorithms for LTE using weights. Paper presented at the 2nd international conference on Applied and Theoretical Computing and Communication Technology (iCATccT), Bengaluru, India, 21–23 July 2016
10. Héliot F, Imran M, Tafazolli R (2013) Low-complexity energy-efficient resource allocation for the downlink of cellular systems. IEEE Trans Commun 61(6):2271–2281
11. Gavrilovska L, Talevski D (2011) Novel scheduling algorithms for LTE downlink transmission. Paper presented at the 19th Telecommunications Forum (TELFOR), Belgrade, Serbia, 22–24 November 2011
12. Kwan R, Leung C, Zhang J (2009) Proportional fair multiuser scheduling in LTE. IEEE Signal Process Lett 16(6):461–464
13. Kaneko M, Popovski P, Dahl J (2006) Proportional fairness in multi-carrier system: upper bound and approximation algorithms. IEEE Commun Lett 10(6):462–464
14. Yang T, Héliot F, Heng C (2015) A survey of green scheduling schemes for homogeneous and heterogeneous cellular networks. IEEE Commun Mag 53(11):175–181
15. Dahlman E, Parkvall S, Sköld J (2011) 4G LTE/LTE-advanced for mobile broadband. Academic Press, Elsevier, Tokyo
16. Sesia S, Toufik I, Matthew PJ (2011) The UMTS long term evolution: from theory to practice. Wiley, London
17. Mehaseb MA, Gadallah Y, Elhamy A, Elhennawy H (2016) Classification of LTE uplink scheduling techniques: an M2 M perspective. IEEE Comm Surv Tutorials 18(2):1310–1335
18. Góra J (2014) QoS-aware resource management for LTE-Advanced relay-enhanced network. Dig J Wireless Commun Networking. https://doi.org/10.1186/1687-1499-2014-178
19. Castañeda E, Silva A, Gameiro A, Kountouris M (2017) An overview on resource allocation techniques for multi-user MIMO systems. IEEE Comm Surv Tutorials 19(1):239–284
20. Clerckx B, Joudeh H, Hao C, Dai M, Rassouli B (2016) Rate splitting for MIMO wireless networks: a promising PHY-layer strategy for LTE evolution. IEEE Commun Mag 54(5): 98–105
21. Jo M, Maksymyuk T, Batista R, Maciel TF, de Almeida A, Klymash M (2014) A survey of converging solutions for heterogeneous mobile networks. IEEE Wirel Commun 21(6):54–62
22. Hossain E, Rasti M, Tabassum H, Abdelnasser A (2014) Evolution towards 5G multi-tier cellular wireless networks: An interference management perspective. IEEE Wireless Commun 21(3):118–127
23. Gesbert D et al (2010) Multi-cell MIMO cooperative networks: a new look at interference. IEEE J Sel Areas Commun 28(9):1380–1408
24. Salman EH, Noordin NK, Hashim SJ, Hashim F, Ng CK (2017) An overview of spectrum techniques for cognitive LTE and LTE-A radio system. Telecommun Syst 65:215–228
25. Liu L et al (2012) Downlink MIMO in LTE-advanced: SU-MIMO vs. MU-MIMO. IEEE Commun Mag 50(2):140–147
26. Thakur R, Kotagi VJ, Murthy SR (2017) Resource allocation and cell selection framework for LTE-Unlicensed femtocell networks. Comput Netw. https://doi.org/10.1016/j.comnet.2017.10.004

27. Bhamri A, Hooli K, Lunttila T (2016) Massive carrier aggregation in LTE-Advanced Pro: impact on uplink control information and corresponding enhancements. IEEE Commun Mag 54(5):92–97

28. Kong C, Peng IH (2017) Tradeoff design of radio resource scheduling for power and spectrum utilizations in LTE uplink systems. J Netw Comput Appl 78:116–124

29. Ohta Y, Nakamura M, Kawasaki Y, Ode T (2016) Controlling TCP ACK transmission for throughput improvement in LTE-Advanced Pro. Paper presented at the Conference on Standards for Communications and Networking (CSCN), Berlin, Germany, 31 October–2 November 2016

30. Pedersen KI, Kolding TE, Frederiksen F, Kovács IZ, Laselva D, Mogensen PE (2009) An overview of downlink radio resource management for UTRAN long-term evolution. IEEE Commun Mag 47(7):86–93

31. Olwal T, Djouani K, Kurien AM (2016) A survey of resource management toward 5G radio access networks. IEEE Commun Surv Tutorials 18(3):1656–1686

32. Li Y, Gao Z, Huang L, Du X, Guizani M (2017) Resource management for future mobile networks: architecture and technologies. Comput Netw. https://doi.org/10.1016/j.comnet.2017.04.007

33. Ahmad A, Ahmad S, Rehmani M, Hassan N (2015) A survey on radio resource allocation in cognitive radio sensor networks. IEEE Commun Surv Tutorials 17(2):888–917

34. 3rd Generation Partnership Project; Technical Specification Group Services and System Aspects; Policy and charging control architecture (Release 12), Release 12, 2014

35. Holma H, Toskala A (2010) Packet Scheduling In: Wigard J, Holma H, Cury R, Madsen N, Frederiksen F, Kristensson M (ed) WCDMA for UMTS: HSPA Evolution and LTE, Fifth edition, Wiley, Chichester, UK, pp 255–291

36. Mishra A, Venkitasubramaniam P (2016) Anonymity and fairness in packet scheduling: a quantitative tradeoff. IEEE/ACM Trans Networking 24(2):688–702

37. Maia AM, Vieira D, Castro MF, Ghamri-Doudane Y (2016) A fair QoS-aware dynamic LTE scheduler for machine-to-machine communication. Comput Commun 89–90:75–86

38. Kumar S, Sarkar A, Sriram S, Sur A (2015) A three level LTE downlink scheduling framework for RT VBR traffic. Comput Netw 91:654–674

39. Blake S, Black D, Carlson M, Davies E, Wang Z, Weiss W (1988) An architecture for differentiated services. RFC 2475, ACM Digital Library

40. Benchaabene Y, Boujnah N, Zarai F (2016) Performance comparison of packet scheduling algorithms for voice over IP in LTE cellular network. Paper presented at the 4th international conference on Control Engineering & Information Technology (CEIT), Hammamet, Tunisia, 16–18 December 2016

41. Subramanian R, Sandrasegaran K, Kong X (2016) Performance comparison of packet scheduling algorithms in LTE-A HetNets. Paper presented at the 22nd Asia-Pacific conference on Communications (APCC), Yogyakarta, Indonesia, 25–27 August 2016

42. Gong Y, Yan B, Lin S, Li Y, Guan L (2016) Priority-based LTE down-link packet scheduling for Smart Grid communication. Paper presented at the 2nd IEEE international conference on Computer and Communications (ICCC), Chengdu, China, 14–17 October 2016

43. Sharifian A, Schoenen R, Yanikomeroglu H (2016) Joint realtime and nonrealtime flows packet scheduling and resource block allocation in wireless OFDMA networks. IEEE Trans Veh Technol 65(4):2589–2607

44. Zander-Nowicka J, Xiong X, Schieferdecker I (2008) Systematic test data generation for embedded software. Paper presented at the international conference on Software Engineering Research & Practice (SERP), Las Vegas, Nevada, USA, 14–17 July 2008

45. Schäuffele J, Zurawka T (2010) Automotive software engineering. Springer, Germany

46. Utting M, Legeard B (2007) Practical model-based testing: A tools approach. Morgan Kaufmann Publishers Inc., San Francisco, CA, USA

47. Russ M, Danzer B, Korotkiy D (2006) Virtueller Funktionstest für eingebettete Systeme: Frühzeitige Fehlererkennung reduziert kostenintensive Iterationszyklen

48. Jondral FK, Schwall M, Nagel S (2011) Model-based waveform design for heterogeneous SDR Platforms with Simulink. Paper presented at the symposium & wireless summer school. Virginia Tech, Blacksburg, USA
49. Haykin S (2005) Cognitive radio: brain-empowered wireless communications. IEEE J Sel Areas Commun 23(2):201–220
50. Farhan M, Naghmash MS, Abbas F (2014) Optimal design for software defined radio based FPGA. J Eng Dev 18(3):148–161
51. Lyrtech Inc (2009) Small form factor SDR evaluation module/development platform. User's guide, 2nd edn. Lyrtech Inc
52. Fodor G, Racz A, Reider N, Temesvary A (2007) Chapter 4: architecture and protocol support for radio resource management (RRM). Long Term Evolution 3GPP LTE radio and cellular technology (Furht B, Ahson SA)
53. Overview of 3GPP Release 13 V0.0.5 (2014) Overview of 3GPP Release 13 V0.0.5
54. Cavalcanti F, Anderson S (2009) Optimizing wireless communication systems. Springer, USA
55. Jang J, Bok L (2003) Transmit power adaptation for multiuser OFDM systems. IEEE J Select Areas Commun 21(2):171–178
56. Holtzman JM (2000) CDMA forward link waterfilling power control. Paper presented at the 51st IEEE vehicular technology conference, Tokyo, Japan, 15–18 May 2000
57. Viswanath P, Tse D, Laroia R (2002) Opportunistic beamforming using dumb antennas. IEEE Trans Inform Theory 48(6):1277–1294
58. Jalali A, Padovani R, Pankaj R (2000) Data throughput of CDMA-HDR a high efficiency-high data rate personal communication wireless system. Paper presented at the 51st IEEE vehicular technology conference, Tokyo, Japan, 15–18 May 2000
59. Kim H, Han Y (2005) A proportional fair scheduling for multicarrier transmission systems. IEEE Commun Letters 9(3):210–212

User Location Forecasting Based on Collective Preferences

Jorge Alvarez-Lozano, J. Antonio García-Macías and Edgar Chávez

Abstract With the proliferation of mobile devices and the huge variety of sensors they incorporate, it is possible to register the user location on the move. Based on historical records, it is feasible to predict user location in space or space and time. Studies show that user mobility patterns have a high degree of repetition and this regularity has been exploited to forecast the next location of the user. Furthermore, proposals have been made to forecast user location in space and time; in particular, we present a spatio-temporal prediction model that we developed to forecast user location in a medium-term with good accuracy results. After explaining how collaborative filtering (CF) works, we explore the feasibility of using collective preferences to avoid missing POIs and therefore increase the prediction accuracy. To test the performance of the method based on CF, we compare our spatio-temporal prediction model with and without using the method based on CF.

1 Introduction

The ability to forecast user location in an accurate way is central to many research areas such as urban planning, healthcare, pervasive and ubiquitous systems, computer networks and recommender systems, location-based services (LBS), to name a few. With the increasing proliferation of mobile devices and the huge variety of sensors incorporated on them, it is possible to register the user location on the move and hence mobile devices become a very rich source of contextual data.

Also, with the recent proliferation of LBS, it is possible to collect a larger amount of location data (discrete) through checkins, geotagged activities, pictures, and so on.

J. Alvarez-Lozano · J. A. García-Macías (✉) · E. Chávez
Computer Science Department, CICESE Research Center, Carr. Ensenada-Tijuana 3918, Ensenada, Mexico
e-mail: jagm@cicese.edu.mx

J. Alvarez-Lozano
e-mail: jalvarez@cicese.edu.mx

© Springer International Publishing AG, part of Springer Nature 2018
M. A. Sanchez et al. (eds.), *Computer Science and Engineering—Theory and Applications*, Studies in Systems, Decision and Control 143,
https://doi.org/10.1007/978-3-319-74060-7_13

Based on historical records, is feasible to predict user location in space or space and time. User mobility patterns have been studied, and researchers have found that people exhibit a high degree of repetition, visiting regular places during their daily activities [1]. This regularity of the past movements has been exploited to forecast the next location of the user. Meanwhile, other researchers have proposed model to forecast user location in space and time [2, 3, 4]. Specifically, in a previous work [4], we presented a spatio-temporal prediction model to forecast user location in a medium-term. For example, assume current time is 10:00 AM (T), what we want to know is: *Where a user will be in the next 3 to 5 h* ($[T, T + \Delta(T)]$ for some $\Delta(T)$). We evaluated our model with realistic data, and we obtained a good accuracy obtained of up to 81.75% for a prediction period of 30 min, and 66.25% considering 7 h. However, it was not possible to obtain a good accuracy for all users. The major cause was the *lack of location data*.

1.1 Collecting Location Data Issues

Although currently it is possible to collect discrete and continuous location data from different sources, there are some issues related to the collect process. In a realistic way, is not always possible to collect location data due to some factors:

- Technology/infrastructure issues. Nowadays, a great amount of locations have Wi-Fi (e.g. home, office, public places). However, users do not have access to all of these access points. Therefore, user location is only partially known by considering connections to access points. Using GPS to collect data also has its disadvantages, the lack of functionality in indoors, and the atmospheric conditions avoid collecting user mobility data.
- Battery issue. Sensing user location through different technologies reduces the battery life to a couple of hours. Also, considering that mobile devices serve different purposes, users manage battery use. Therefore, it is only possible to sensing user mobility for a limited time period.
- Privacy. Another important issue is user privacy. Not all users are willing to provide their location data; explicitly, users can block the location data collecting process, either at defined time periods, or permanently.

Due to these issues is not possible to have a complete understanding of user mobility, and thereby the user mobility modeling and the prediction model definition would be inaccurate. As most of the forecasting algorithms take as reference a set of significant places or points of interest (POIs) to realize user location prediction, due to the above issues, it is not possible to identify POIs (and associated information) and consequently the prediction models do not represent user mobility on an accurate way; predictions are not accurate. Thus, the question that arises is *how to compensate the lack of location data to avoid missing point of interest* and define a better prediction model. The hypothesis of this paper is that is possible to avoid the lack of mobility data (specifically avoid missing POIs) by considering the mobility of users who are similar to a given user.

1.2 Collective Preferences Rule Our Lives

Generally, our decisions are influenced by the preferences of others; we buy a product because our acquaintances also bought it; we watch a movie because our friends watched it. At the same way, we influence to other people. Therefore, there are some people who are similar considering the products they buy, the movies they watch and so on. This aspect also applies for the places we visit; we visit a restaurant because a friend invited us; we visit a park because some friends do exercise there. Besides that, we are similar to other users because we have some places in common (due to scholar, work, or leisure activities). Therefore, to have knowledge about what places have been visited for a given user, is feasible to consider the places visited by his/her similar users, and hence avoid missing that some places are considered POIs. To do this it is feasible to use a recommender system technique, *collaborative filtering*.

Collaborative Filtering (CF) is a technology that has emerged in e-Commerce applications to produce personalized recommendations for users [5]. It is based on the assumption that people who like the same things are likely to feel similarly towards other things. This has turned out to be a very effective way of identifying new products for customers. CF works by combining the opinions of people who have expressed inclinations similar to a given user in the past to make a prediction on what may be of interest to him right now. One well-known example of a CF system is Amazon.com.

In this work, we present a spatio-temporal prediction model that incorporates a method based on collaborative filtering to avoid missing points of interest (and associated data), and consequently we define a better prediction model. To know the performance of the proposed approach, we compare the results with those obtained by our prediction model presented in [4]. The rest of this paper is organized as follows: we start discussing the characteristics of the mobility that allow us to model mobility, after that we present on a general way our previous spatio-temporal prediction model. Later, we describe how collaborative filtering works. Then, in Sect. 5, we describe how avoid missing POIs by considering the places that have been visited by similar users. After that, we present the evaluation and results.

2 User Mobility

Although user mobility seems to be dynamic, most people follow certain mobility patterns [1, 6, 7]; it is rare to have a completely erratic behavior over time. Mobility is fixed by our activities and habits, like working, school attendance, recreational

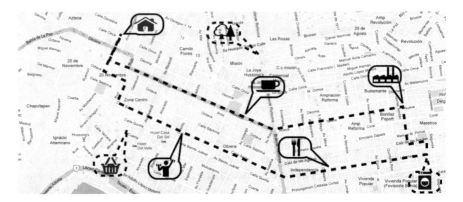

Fig. 1 Some significant places are found in the daily user activities

endeavors, and other activities that vary over time within certain behavior boundaries. We can distinguish between weekday, weekend, monthly or annually patterns [6, 8, 9]. Once recognized the user mobility patterns, we are able to predict her spatiotemporal location. In our previous work [4], we considered some features and their interrelation to model user mobility (Fig. 2 illustrates these features). It is reasonable to assume certain periodicity in location/time patterns. Usually weekdays are similar, people tend to organize their life according to work or school hours; activities in the same weekday will have a repetitive pattern. A corresponding periodicity is observed during weekends. Mobility exhibits a different pattern for each day of the week [7, 8, 10, 11]. Hence, the places visited in a given day are postulated to be the same for the subsequent days (*feature: day of the week*) (Fig. 1).

Two additional features are that the current location in a given day and hour conditions the next place to be visited. For example if one user is at home at 7:00 AM on a Monday (*feature: hour of the day*), the next place he will be at is most likely the coffee shop or the office, but not the restaurant or a movie theater (*feature: current location*). We postulate the user mobility is a markovian stochastic process

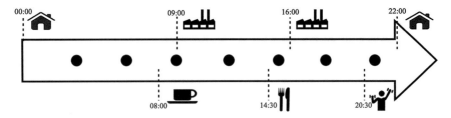

Fig. 2 Some The next user location depends only on the current location, once the sequences have been factorized by week day and time of the day

and can be described with a Markov Chain. The Markovian property [12] would state that current place is only a function of the previous place. Our main claim is that once the data is grouped by day and time, the sequence of places visited form a Markovian chain.

2.1 Markovian Chain Among POIs

It is necessary to clarify that the markovian property is just valid when the user moves among certain discrete places. These discrete locations are distinguished because the user spends some time and visit them frequently (see Fig. 1); they are known as significant places or points of interest (POI). In our previous work, we found that the residence at these POIs is related at the time of the day. Therefore, considering the Markovian property and the relationship of the residence at POIs with time of the day, was feasible to model user mobility as a hidden Markov Model (see Fig. 3). In the next section we describe our prediction model in a general way.

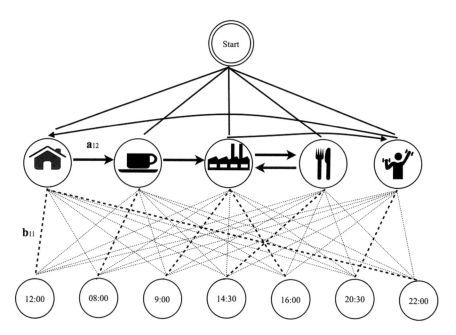

Fig. 3 An HMM representing some POIs and their relationship with different hours of the day

3   Spatio-temporal Prediction Model

3.1   Modeling User Mobility as a Hidden Markov Model

A hidden Markov model is a finite state machine consisting of a set of hidden states (Q), a set of observations (O), transition probabilities (A), emission probabilities (B), and initial probabilities for each state (π). Hidden states are not directly visible, while observations (dependents on the state) are visible. Each state has a probability distribution over the set of observations ($\lambda = (A; B; \pi)$). In our case, the hidden states correspond to POIs, which have a probability distribution over times of day. Once the HMM has been defined, we use the *decoding approach*; given a time period (sequence of observations), we want to know the most likely sequence of locations where the user will be (hidden states).

3.2   Defining User Prediction Model

In our previous approach, the HMM is defined as follows (see Fig. 3):

- Hidden states. These are defined by the set of POIs. Also, another hidden state was added to define when the user is at a location that is not a POI.
- Observations. These are defined by the average of the arrival and leaving time to the POIs. According to [13], the arrival and leaving times to some places do not change much. This way, we can define in an accurate way the time when the user will be at a POI; otherwise, considering the leaving time, we can define that he is at another POI, or at a non POI.
- Vector π. It defines the probability that the user starts his day at a given POI.
- Transition matrix. It defines the probability that the user moves from a POI to another, or from a POI to the state that corresponds to a non-POI.
- Confusion matrix. It defines the probability that the user is at a given POI (or at a non-POI), at a given time.

Regarding to discover the hidden states sequence that was most likely to have produced a given observation sequence, we use the Viterbi's algorithm [14]. Once the HMM is defined, we are able to forecast the user location in a given time period (see Fig. 4). For instance, if the current time is 11:00 AM, and we want to know where the user will be in the period 11:00–18:00, the vector π, transition matrix (A), and confusion matrix (B), are used to identify the combinations of hidden states (and their corresponding probabilities) that satisfy such time period, so later the sequence of hidden states with the highest probability is selected. In order to define the prediction model, some stages are required, here we summarized these stages:

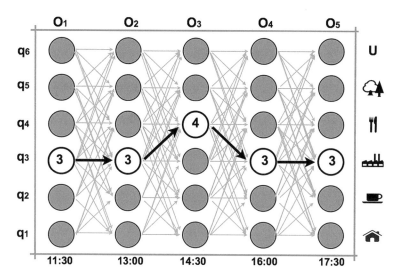

Fig. 4 Using Viterbi's algorithm to identify the sequence of POIs where the user will be in a given time period

3.3 Identifying Points of Interest

Mobile traces produced by the mobile devices, provide a great amount of location data useful to discover where the user spends her time. Thereby, with this amount of location data there is a need for algorithms that deal with the challenge of turning data into significant places [15–22].

To discover POIs, we have used the algorithms presented by Ashbrook and Starner [15] and Kang et al. [16] taking into account the below criterions:

- Residence time.
- Cluster radius.
- Frequency of visits.
- Time period (Windows size).

The Ashbrook and Starner's [15] algorithm focuses on discovering indoor significant places. To do that, they consider the GPS loss signal within a fixed radius r, and a time threshold t for the disappearing period. Meanwhile, Kang et al. [16] propose a time-based clustering algorithm to discover outdoor significant places. They compare each incoming GPS reading with previous readings in the current cluster; if the stream of readings moves away from the current cluster, then they form a new one. In Kang et al. [16] consider two thresholds, d and t for distance and staying time respectively. If the GPS readings are close together (within some distance d of each other), and the user spends at least t minutes at that region, a cluster is formed. When the user moves away from the current cluster, a new cluster is formed; the cluster is discarded if the user stays less than t time. These variables allow us to identify the places where the user spends some time and visits frequently. Also, considering the cluster radius we can identify POIs in different levels. The last variable is very important to define an accurate forecasting model, and it is necessary to identify the period of time that covers the current mobility pattern. That is, if the current day is a Monday i, the challenge is to determine the quantity of previous Mondays that are similar in terms of mobility.

3.4 User Mobility Similarity

To identify the time period that covers the current mobility pattern, we used the cosine similarity; comparing the Day_i vector with the Day_{i-1} vector. The window size increases if the similarity is above some threshold Θ; otherwise, skip the records of the Day_{i-1}, and compare Day_i vector and Day_{i-2} vector. If the similarity of m consecutive days is below some threshold Θ, the window size ends, and just includes the records of the days with a similar mobility.

3.5 Converting User Mobility into a Vector

In order to compare the user mobility by day, each day has been converted into a vector. For each day, we divide it into 48 periods of 30 min; each slot contains an index (starting from 1) that corresponds to a place where the user has been in that period, as shown in Fig. 5. The index 0 defines an *unknown place*. We defined the size of the slot to 30 min in order to know with this level of granularity where the user has been.

Week \ Time	...	07:30	08:00	08:30	09:00	...	23:30
1	...	1	0	0	2		1
2	...	0	0	0	2		1
3	...	0	0	2	2		1

Fig. 5 User mobility as a vector

3.6   Updating POIs

Also, in our previous work, we incorporated a mechanism to update POIs (and the information related to them) day by day. Considering that user preferences or interests change over time, it is necessary to include new POIs to the prediction model. The prediction model also is updated when a place ceases to be a POI.

3.7   Predictability of the User Mobility

And finally, as our approach is based on the hypothesis that the user mobility among POIs can be represented as a Markov chain, it is important to verify that the user mobility has the Markov property in order to do an accurate prediction; otherwise, HMM is not useful, and it will generate bad results. To address this aspect, we use the test proposed in Zhang et al. [23]. This test uses the transition frequency matrix to determine whether user mobility has the Markov property. More detail about the spatio-temporal prediction model can be found in Alvarez-Lozano et al. [4].

4   Collaborative Filtering

Collaborative filtering (CF) is a popular recommendation algorithm that bases its predictions and recommendations on the ratings or behavior of other users in the system. The fundamental assumption behind this method is that other users' opinions can be selected and aggregated in such a way as to provide a reasonable prediction of the active user's preference. Intuitively, they assume that, if users agree about the quality or relevance of some items, then they will likely agree about other items—if a group of users likes the same things as John, then John is likely to like the things they like which he hasn't yet seen [24]. Examples of its use include Amazon, iTunes, Netflix, LastFM, StumbleUpon, and Delicious.

The information domain for a collaborative filtering system consists of users which have expressed preferences for various items. A preference expressed by a user for an item is called a rating and is frequently represented as a (User, Item, Rating) triple. These ratings can take many forms, depending on the system in question.

The set of all rating triples forms a sparse matrix referred to as the ratings matrix. The rating matrix is denoted by $\mathbf{R}$, with $r_{u,i}$ being the rating user u provided for item i, r_u being the vector of all ratings provided by user u, and r_i being the vector of all

	I₁	I₂	I₃	I₄	I₅	I₆	I₇	I₈	Iᵢ	I_N
U₁	$r_{1,1}$	$r_{1,2}$	$r_{1,3}$						...	$r_{1,N}$
U₂									...	
U₃									...	
U₄									...	
U₅									...	
U₆									...	
U₇									...	
U₈	$r_{8,1}$								...	
Uⱼ	...	...	...	...	...	...	...	...	...	...
U_M	$r_{M,1}$								...	$r_{M,N}$

Fig. 6 Rating matrix presenting the ratings that M users have regarding to N items

ratings provided for item i. Figure 6 presents an example of the rating matrix. Generally, CF can be categorized as user-user and item-item.

4.1 User-User Collaborative Filtering

User-user CF is a straightforward algorithmic interpretation of the core premise of collaborative filtering: find other users whose past rating behavior is similar to that of the current user and use their ratings on other items to predict what the current user will like. To predict John's preference for an item he has not rated, user-user CF looks for other users who have high agreement with John on the items they have both rated. These users' ratings for the item in question are then weighted by their level of agreement with John's ratings to predict John's preference.

4.2 Item-Item Collaborative Filtering

Item-item collaborative filtering, also called item-based collaborative filtering is one of the most widely deployed collaborative filtering techniques today [25, 26]. Amazon is a good example of the usage of this technique. Rather than using similarities between users' rating behavior to predict preferences, item-item CF uses similarities between the rating patterns of items. If two items tend to have the same users like and dislike them, then they are similar and users are expected to have similar preferences for similar items.

4.3   Stages of CF

For any approach the stages of the CF are as follows:

4.3.1   Building a User Profile

The first stage is to build user profiles from feedback on items made over time. A user profile comprises these numerical ratings assigned to individual items. More formally, each user u has at most one rating $r_{u,i}$ for each item i.

4.3.2   Measuring User Similarity

In user-based CF, the goal is to locate other users with profiles similar to that of the active user, commonly referred to as *neighbors*. This is done by calculating the weight of the active user against every other user with respect to the similarity in their ratings given to the same items. For item-based CF, the goal is to identify items that are similar to a given item.

4.3.3   Generating a Prediction

After similar users or items are identified, it is possible to combine all the neighbors' ratings into a prediction by computing a weighted average of the ratings, using the correlations as the weights.

For user-based CF, the system combines the ratings of users in N to generate predictions for user u's preference for an item i. This is typically done by computing the weighted average of the neighboring users' ratings i using similarity as the weights [27]:

$$p_{u,i} = \overline{r_u} + k \sum_{v=1}^{n} \left(\overline{r}_{v,i} - \overline{r}_v \right) \cdot w_{u,v} \qquad (1)$$

where n is the number of best neighbors chosen and k is a normalizing factor such that the absolute values of the weights sum to unity.

For item-based CF, after collecting a set S of items similar to i, $p_{u,i}$ can be predicted as follows:

$$p_{u,i} = \frac{\sum_{j \in S} s(i,j) r_{u,j}}{\sum_{j \in S} |s(i,j)|} \qquad (2)$$

S is typically selected to be the k items most similar to i that u has also rated for some neighborhood size k.

5 User Location and CF

Considering the POIs identification process, in this work (and in most of the related works), we use a minimum number of visits (frequency) to consider a place as a POI. However, this decision has a problem. Consider that 3 users, John, Mark, and Sam are coworkers, and during the workday, they have in common some places such as coffee shop, restaurant, meeting room, and an office. However, in some occasions John's mobile device had no battery, and it was not possible to collect location data while he was in the restaurant; the restaurant does not have the required visits to be considered as a POI. In contrast, Mark and Sam had no problem to collect location data while they were in the restaurant. For these users the restaurant was considered as a POI. But considering that the three users have some places in common, and they differ just in one place (*restaurant*), there is some probability that the restaurant has to be considered as a POI for John.

Hence, unlike the approach of collaborative filtering to make predictions for a user considering the preferences of users similar to this one, in this paper we are interested in avoid missing POIs by considering the places that are visited by those users who are similar to a given user. Therefore, instead of having items' ratings, we have places' ratings. Thus, in the R matrix, the rows represent *users* and columns represent geographic regions. Hence, we select the *geographic region* where user of interest mobilize.

5.1 Building User Profile

Once we select the geographic region, we proceed to divide it into $N \times M$ cells of a given size as shown in Fig. 7a, each cell represents a column in the R matrix (cell 1, 1 represents column 1; cell 1, 2 represents column 2; cell N, M represents column $N \times M$). After that, in order to define the vector r_u for a user i, we take as reference the geographical coordinates of user i' POIs. The r_u vector indices are marked with 1 to indicate that in the correspondent cell there is a POI. Otherwise, the vector indices are marked with 0. For example, considering the Fig. 7, a user has 6 POIs (Fig. 7a), the correspondent cells are marked in Fig. 7b, and finally, the correspondent indices are marked in the r_u vector (Fig. 7c). Thus, we have knowledge about in which cells the user spends his time. After defining each vector r_u, the matrix R is constructed (Fig. 8), and at this point it is possible to compute the similarity among each pair of users.

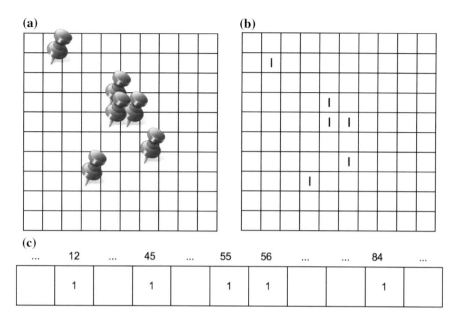

Fig. 7 Building user profile **a** The geographic region is divided into cells of a given size. **b** Cells are marked according to the POIs geographic position. **c** The correspondent indices in the r_u vector are marked

Fig. 8 Once the matrix R is defined, it is possible to compute the similarity among users

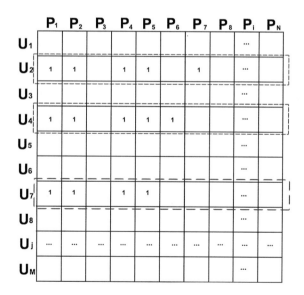

5.2 Measuring User Similarity

Considering the mentioned scenario, John, Mark and Sam vectors can be defined as r_{u7}, r_{u4}, and r_{u2} (see Fig. 8), and using a similarity function, a certain similarity is found between John and Mark, and John and Sam. There are different mathematical formulations that can be used to calculate the similarity between two items. Using cosine similarity, users are represented as vectors and similarity is measured by the cosine distance between two rating vectors:

$$sim(u, v) = \cos(\vec{u}, \vec{v}) = \frac{\vec{u} \cdot \vec{v}}{\|\vec{u}\| * \|\vec{v}\|} \tag{3}$$

5.3 Avoiding Missing Points of Interest

After we know the similarity among users, for each user we select the k more similar users to this one.

Thus, to avoid missing POIs we consider the similarity between r_U and r_V (V is the most u' similar vectors) vectors, if similarity is greater than a given threshold θ, we compare the indices marked as 1 in both vectors, if vector r_V has a larger amount of marked indices, we verified that user i on day of the week j associated to vector r_U has location records (candidate POIs) in those regions in which differs with vector r_v.

If user i on the day of the week j has at least n visits to the candidate POIs, and the visits to these places were realized during the time period that comprises the current mobility pattern, these places are considered as POIs. In order to know the performance of the prediction model after adding new POIs, a new prediction model is defined when new POIs are added. This process is applied considering each k value. The process of incorporating POIs is restricted by the date in which visits to the candidates POIs were realized. That is, if we add a place that was visited outside of the current mobility pattern, the prediction model does not define the user mobility behavior in an accurate way, and the predictions made by this prediction model will be inaccurate.

The above process is applied for each vector r_u. Thus, for each user and day of the week, the incorporation of places (candidate POIs) as POIs is verified considering the places visited by the k most similar users.

It is important to remark that since POIs are identified for each day of the week, and therefore a prediction model is defined for each day of the week, we define a vector for each day. Thus, we define seven vectors for each user; r_{id}, represents the cells that user i has visited in the day of the week d.

6   Evaluation

As the main goal of this paper is to know the feasibility of using collective preferences to avoid missing POIs and therefore increase the prediction accuracy, to test the performance of the method based on CF, we compared our spatio-temporal prediction model with and without using the method based on CF.

6.1   *Dataset*

To evaluate our approach we use the GeoLife dataset [21, 28, 29], which contains GPS trajectories collected in the context of the GeoLife project from Microsoft Research Asia. This dataset contains trajectories of 178 users collected in a period of four years, from April 2007 to October 2011. A GPS trajectory is represented by a sequence of time-stamped points, each one of them containing the information of latitude, longitude, and altitude. The trajectories were recorded by different GPS loggers and GPS-phones, and have a variety of sampling rates, with 91% percent of the trajectories being logged in a dense representation (every 1–5 s or every 5–10 m per reading). After analyzing the records of each user, we decided to choose the records of 35 users because these users have continuous GPS readings over several weeks, allowing us to define an accurate model for them. Also, these users realize their daily activities in the same city: Beijing, China.

6.2   *Training Prediction Models*

For each user, we grouped the records according to the day when they were created. Then, we used the readings of the last month to test the prediction model, and the remaining readings are used to identify the accurate time period for training the prediction model. This way, we use the historical records of a given day of the week to define the prediction model for this specific day. For each user, seven spatio-temporal prediction models were defined in order to characterize the user mobility by day. Once the time period that includes the current mobility pattern has been identified, the prediction model training is realized. After that, the process that determines the incorporation of new POIs is performed. In order to determine the viability and effectiveness of the incorporation of new POIs, for each value of K considered, a new prediction model is defined if, and only if new places were added to the initial set of POIs. This way, for each day of the week, each users has up to 9 predictions models. That is, for each size of POI, the user has the base prediction model and the resulting prediction models after considering the similarity with the 1, and 3 most similar vectors.

6.3 Defining User Profile

In order to define each vector r_u, first we define the geographic region. As the users considered live in Beijing, we select the area of this city, as shown in Fig. 9a. As the algorithms to identify POIs uses three values to set the cluster radius (500, 250, and 100 m), we divide the geographic region into cells considering three values: 1000, 500, and 250 m (see Fig. 9b, c and d, respectively). Thus, we define three matrixes R: R_{1000}, R_{500}, and R_{250}, allowing us to know the user similarity at three different levels. Hence, we define 21 vectors for each user; 7 for each matrix R. This way, we use the POIs with radius of 100 m to define the vectors of the matrix R_{250}, the POIs with radius of 250 m to define the R_{500}, and finally the POIS with radius of 500 m to define the R_{1000}.

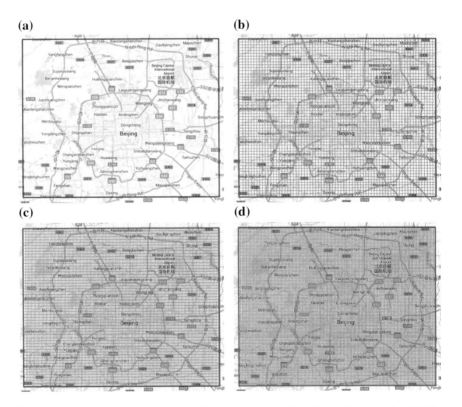

Fig. 9 Geographic region used to define R matrix **a** Geographic grid considering cells of 1000 m **b** Geographic grid considering cells of 500 m **c** Geographic grid considering cells of 250 m

6.4 Predictions

For each user, there is a month of records for the test process. This way, each prediction model can be tested 4 on four days. Thus, we used each prediction model (i.e. base, $k = 1$, $k = 3$, as appropriate) to predict user location for the correspondent day in the test week 1. Once we did predictions for test week 1, we compare the user mobility of the predicted day with the mobility of the training process using the cosine similarity. If the similarity is above some threshold Θ ($\Theta = 0{:}50$), the current HMM is updated with the records of the predicted day. Otherwise, the current HMM is used to predict the user location for the test week 2 (i.e. 9th Monday). This process is applied for the subsequent weeks 3 and 4. For each spatio-temporal prediction model, we have made five predictions considering different values for ΔT (30 min, 1, 3, 5, and 7 h): 20 predictions for each prediction model. All the prediction models were defined using a 1st order HMM.

6.5 Effectiveness of the Prediction Model

To determine the effectiveness of the prediction, if we estimate where a user will be in the interval $[T; T + \Delta T]$, the prediction is correct if the user is at place q_i in the interval $[T_{pred} - \theta; T_{pred} + \theta]$; θ represents an error margin. That is, a prediction is correct when the user is at the POI defined by q_i, at the time indicated by the observation o_i with certain error margin. It would also be correct if the prediction indicates that the user is not at a POI (in the case of the state corresponding to an unknown place). We have defined $\theta = 15$ min.

$$T_{pred} = T + o_i \quad 1 \le i \le \text{No. of obs. in the interval} \tag{4}$$

7 Results

7.1 POIs

Regarding to points of interest, on average each user has 3.65 POIs when we set the cluster radius to 500 m; 3.62 POIs using a cluster radius of 250 m, and 3.20 POIS with a cluster radius of 100 m (Table 1).

Table 1 Average number of POIs identified per user

Cluster radius	POIs
500 m	3.65
250 m	3.62
100 m	3.20

7.2 Matrix R Vectors r_u

Once we have identified POIS, the vectors r_u were defined. Table 2 presents the amount of vectors defined considering different cell sizes. If for each user and day of the week we define a vector r_u, each matrix R would have 245 vectors. However, not all users have POIs in each day of the week, because that, a lower number was obtained, just 232 vectors for each R matrix.

7.3 Vectors Similarity

Later, for each matrix, we obtained the similarity of each vector compared to the rest of the vectors. To the approach of this paper, becomes relevant those vectors with similarity greater than a given threshold. However, not all vectors have at least one vector with similarity greater than the threshold. Therefore, Table 2 shows the amount of vectors which have at least one vector with a similarity greater than a given threshold; the threshold was set to 0.75. After that, for each vector, we identified the k vectors most similar. Also, Table 2 presents the number of vectors with 3 and 5 similar vectors (similarity > threshold).

Thereafter, Table 3 presents the average similarity for all vectors considering each k value. As can be seen, as the cell size increases, the similarity also increases. Moreover, in Table 4 we present the maximum and minimum similarity (average) found when we take as reference the vectors associated to each day of the week. An interesting aspect is the fact that by considering different cell size and k values, the maximum similarity is obtained when a work day is taken as reference. This can be explained by the fact the users do not have regular mobility patterns on weekends. Another interesting fact is that vectors associated to weekdays have a higher similarity by comparing them with other vectors. For example, when cell size is set to 1000 m, 41% of vectors associated to weekdays have at least one vector with

Table 2 Number of vectors defined using different cell sizes, and number of vector which have at least one vector with a similarity greater than a given threshold

Cell size	Vectors	k = 1	k = 3	k = 5
500 m	232	136	63	28
250 m	232	114	60	38
100 m	232	63	1	0

Table 3 Average similarity found by considering different cell sizes and K values

Cell size	k = 1	k = 3	k = 5
1000	0.8796	0.8727	0.8499
500	0.8715	0.8530	0.8353
250	0.8470	0.8208	

Table 4 Major and minor similarity found by similarity between vectors when considering a vector per week of the week

Size	k = 1	Day	k = 3	Day	k = 5	Day
1000	0.920	Mon	0.900	Mon	0.863	Mon
	0.861	Sat	0.853	Tue	0.838	Tue
500	0.901	Tue	0.867	Mon	0.849	Sun
	0.830	Sun	0.815	Thu	0.802	Thu
250	0.874	Thu	0.826	Fri		
	0.782	Mon	0.786	Mon		

similarity higher than the threshold. In contrast, for vectors associated to weekends, just the 20%. The same behavior is identified when using different cell sizes (see Table 5).

Later, in Table 6 we present another interesting aspect. One problem associated with getting similarity among vectors, as the number of vectors increases when more users are considered; getting vector similarity process is complicated.

Therefore, we identified the amount of occasions on which the k most similar vectors correspond to the same user. For example, considering the cell size of 1000 m and $k = 5$, 28 vectors have 5 similar vectors (similarity $> \Theta$), and on 4 times, these vectors correspond to the user (14.28% of times). Or, considering a cell size of 500 m and $k = 1$, 114 vectors have 1 similar vector, and on 97 times this vector corresponds to the same user (85.08% of times). When the cell size was set to 250 m, just a few amount of vectors have 1 similar vector (63), and on 49 times these vectors correspond to the same user; however, considering $k = 3$ just 1 vector have 3 similar vectors and corresponds to the same user. And finally, with $k = 5$ there is no vector with 5 similar vectors. As cell size decreases, the lower the amount of similar vectors. Considering the above results, it is feasible to obtain just

Table 5 Percentage of vectors having k similar vectors (sim $> \Theta$) according of the type of day: WD: weekday, and WE: weekend

Cell size	k = 1	k = 3	k = 5
1000 WD	41.14	14.85	8.57
1000 WE	24.63	5.79	0
500 WD	44.57	23.42	11.42
500 WE	20.28	10.14	4.34
250 WD	17.33	0.06	0
250 WE	3.38	0	0

Table 6 Amount of times when the K most similar vectors correspond to the same user

Cell size	k = 1	k = 3	k = 5
1000 (No.)	92	31	4
Percentage	67.64	49.20	14.28
500 (No.)	97	25	4
Percentage	85.08	41.66	10.52
250 (No.)	49	1	
Percentage	77.77	100	

the similarity among the vectors associated to a given user, due that by considering different values for k, in most occasions the most similar vectors corresponded to the same user.

7.4 Incorporating POIs

After applying the process to avoiding missing POIs, some additional prediction models were defined. In Table 7, we present the number of models defined considering different cell sizes and $k = 1$ and $k = 3$. For example, considering a cell size of 500 m, of 60 vectors having 3 similar vectors, just on 12 occasions were added POIs, and therefore 12 additional prediction models were defined. When the cell size is of 250 m, of 63 vectors having 1 similar vectors, just were defined 9 additional prediction models. Regarding POIs added, Table 8 presents the results. Considering a cell size of 1000 m and $k = 1$, on average 1.62 were added, in contrast 1.45 were added when k = 3. When the process considered a cell size of 500 m, there were added 1.36 and 1.3 POIs for $k = 1$ and $k = 3$, respectively. And, for a cell size of 250 m and $k = 1$, on average 1.3 POIs were added. As the reader can notice, Table 8, does not show the amount of POIs added when k was set to 5. Although, the process to avoid missing POIs identified places to be added considering $k = 5$, these places had already been added when k was set to 1 or 3. Therefore, we can argue that it is just necessary to consider the most similar vector and the three most similar vectors to realize the POIs incorporation process.

7.5 Prediction

In Fig. 10, we show the average accuracy obtained on the four test weeks by considering the prediction models that used POIs of different radius size. When

Table 7 Extra prediction models defined

Cell size	k = 1	k = 3
1000	49	14
500	33	12
250	9	0

Table 8 Average amount of POIs added after the POIs incorporation process

Cell size	k = 1	k = 3
1000	1.62	1.45
500	1.36	1.30
250	1.30	0

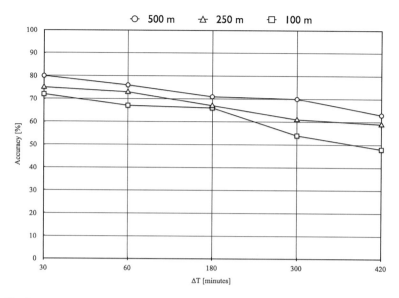

Fig. 10 Once Average accuracy obtained after

considering a period of 30 min, we obtain an accuracy of 80% considering POIs of 500 m; for a period of 60 min, we obtain 76%. When ΔT is set to 180 min, the accuracy is of 71% for a period of 3 h; a period of 5 h yields an accuracy of 70%, and finally a 7 h period yields an accuracy of 63%. When considering POIs of 250 m, the accuracy is of 75, 73, 67, 61, and 59%, respectively; when considering POIs of 100 m, an accuracy of 72, 67, 66, 54, and 48% was obtained. These results correspond to the accuracy obtained by the base prediction models.

Regarding to the accuracy increase obtained by the prediction models defined after the process to avoiding POIs, in Table 9 we present the average increase. The predictions models defined by comparing vectors associated to cells of 1000 m and $k = 1$, obtained an increase of 8.14% over the base prediction model, and 7.97% when $k = 3$. Likewise, when the cell size was set to 500 m, the increase was 7.84 and 8.32% for $k = 1$ and $k = 3$, respectively. And finally, for a cell size of 250 m and $k = 1$, the increase was of 8.45%. It is important to mention that it was obtained an accuracy of up to 13%.

Table 9 Accuracy increase by considering the prediction models defined after considering the most similar vector and the three most similar vectors

Radius cluster	$k = 1$	$k = 3$
1000	8.14	7.97
500	7.84	8.32
250	8.45	0

References

1. Gonzalez MC, Hidalgo CA, Barabasi A-L (2008) Understanding individual human mobility patterns. Nature 453(7196):779–782
2. Scellato S, Musolesi M, Mascolo C, Latora V, Campbell AT (2011) NextPlace: a spatiotemporal prediction framework for pervasive systems. In: Proceedings of the 9th international conference on pervasive computing, Pervasive '11, Springer, pp 152–169
3. Sadilek A, Krumm J (2012) Far out: predicting long-term human mobility. In: Proceedings of the 26th AAAI conference on Artificial Intelligence, AAAI, AAAI Press
4. Alvarez-Lozano J, García-Macías JA, Chávez E (2015) Crowd location forecasting at points of interest. Int J Ad Hoc Ubiquitous Comput 18(4):191–204
5. Schafer, J. B., Konstan, J. & Riedl, J. (1999), Recommender systems in e-commerce. In: Proceedings of the 1st ACM conference on Electronic Commerce, EC '99, ACM, New York, NY, USA, pp 158–166
6. Eagle N, Pentland A (2006) Reality mining: sensing complex social systems. Pers Ubiquit Comput 10(4):255–268
7. Farrahi K, Gatica-Perez D (2011) Discovering routines from large-scale human locations using probabilistic topic models. ACM Trans Intell. Syst Technol 2(1):3:1–3:27
8. Motahari S, Zang H, Reuther P (2012) The impact of temporal factors on mobility patterns. In: Proceedings of the 2012 45th Hawaii International Conference on System Sciences, HICSS '12, IEEE Computer Society, pp 5659–5668
9. Yavas G, Katsaros D, Ulusoy O, Manolopoulos Y (2004) A data mining approach for location prediction in mobile environments. Data & Knowl Eng 54(2005):121–146
10. Chon Y, Shin H, Talipov E, Cha H (2012) Evaluating mobility models for temporal prediction with high-granularity mobility data. In: Proceedings of the 2012 IEEE international conference on Pervasive Computing and Communications, Percom, IEEE, pp 206–212
11. Hsu W, Spyropoulos T, Psounis K, Helmy A (2007) Modeling time-variant user mobility in wireless mobile networks. In: Proceedings of the 26th IEEE international conference on computer communications, INFOCOM, IEEE, Anchorage, Alaska, USA, pp 758–766
12. Markov AA (1961) Theory of algorithms, Israel program for scientific translations. Bloomington, IN, USA
13. Do TMT, Gatica-Perez D (2012) Contextual conditional models for smartphone-based human mobility prediction. In: Proceedings of the 2012 ACM conference on Ubiquitous Computing, UbiComp '12, ACM, pp 163–172
14. Viterbi AJ (2006) A personal history of the Viterbi algorithm. IEEE Signal Process Mag 23 (4):120–142
15. Ashbrook D, Starner T (2003) Using gps to learn significant locations and predict movement across multiple users. Pers Ubiquit Comput 7(5):275–286
16. Kang JH, Welbourne W, Stewart B, Borriello G (2005) Extracting places from traces of locations. SIGMOBILE Mob Comput Commun Rev 9(3):58–68
17. Kim M, Kotz D, Kim S (2006) Extracting a mobility model from real user traces. In: Proceedings of the 25th IEEE international conference on computer communications, INFOCOM '06, pp 1–13
18. Marmasse N, Schmandt C (2000) Location-aware information delivery with commotion. In: Proceedings of the 2nd international symposium on Handheld and Ubiquitous Computing, HUC '00, Springer, pp 157–171
19. Palma AT, Bogorny V, Kuijpers B, Alvares LO (2008) A clustering-based approach for discovering interesting places in trajectories. In: Proceedings of the 2008 ACM Symposium on Applied Computing, SAC '08, ACM, pp 863–868
20. Ram A, Jalal S, Jalal AS, Kumar M (2010) A density based algorithm for discovering density varied clusters in large spatial databases. Int J Comput Appl 3(6):1–4 (Published by Foundation of Computer Science)

21. Zheng Y, Zhang L, Xie X, Ma W-Y (2009) Mining interesting locations and travel sequences from gps trajectories. In: 'Proceedings of the 18th international conference on World Wide Web', WWW '09, ACM, pp 791–800
22. Zhou C, Frankowski D, Ludford P, Shekhar S, Terveen L (2007) Discovering personally meaningful places: an interactive clustering approach. ACM Trans Inf Syst 25(3):56–68
23. Zhang YF, Zhang QF, Yu RH (2010) Markov property of markov chains and its test. In: Proceedings of the International Conference on Machine Learning and Cybernetics (ICMLC), IEEE, pp 1864–1867
24. Ekstrand MD, Riedl JT, Konstan JA (2011) Collaborative filtering recommender systems. Found Trends R in Hum-Comput Interact 4(2):81–173
25. Sarwar B, Karypis G, Konstan J, Riedl J (2001) Item-based collaborative filtering recommendation algorithms. In Proceedings of the 10th international conference on World Wide Web, WWW '01, ACM, New York, NY, USA, pp 285–295
26. Karypis G (2001) Evaluation of item-based top-n recommendation algorithms. In Proceedings of the tenth international Conference on Information and Knowledge Management, CIKM '01, ACM, New York, NY, USA, pp 247–254
27. Breese JS, Heckerman D, Kadie C (1998) Empirical analysis of predictive algorithms for collaborative filtering. In: Proceedings of the fourteenth conference on Uncertainty in Artificial Intelligence, UAI '98, Morgan Kaufmann Publishers Inc., San Francisco, CA, USA, pp 43–52
28. Zheng Y, Li Q, Chen Y, Xie X, Ma W-Y (2008) Understanding mobility based on gps data. In: Proceedings of the 10th international conference on Ubiquitous Computing, UbiComp '08, ACM, pp 312–321
29. Zheng Y, Xie X, Ma W-Y (2010) Geolife: a collaborative social networking service among user, location and trajectory. IEEE Data Eng Bull 33(2):32–39

Unimodular Sequences with Low Complementary Autocorrelation Properties

Israel Alejandro Arriaga-Trejo and Aldo Gustavo Orozco-Lugo

Abstract The design of sequences with constant amplitude in the time domain, that possess specific autocorrelation and specific complementary autocorrelation functions, is here addressed. The proposed sequences can be used in the identification of strictly linear (SL) and widely linear systems (WL) by making use of the second order characteristics of the process observed at the output of the system. Through the analysis developed, the main differences of the disclosed sequences with those commonly used for conventional SL processing are highlighted. Theoretical results are accompanied with numerical simulations to show the performance of the sequences here designed.

1 Introduction

Sequences with constant amplitude in the time domain, whose autocorrelation function is zero except at some correlation lags, have been the subject of interest in the research community since the 1960s [1–3]. This is due to their applications in different fields of knowledge such as channel estimation and access codes in wireless communications systems [4], active sensing, medical imaging, radar waveform design [5], and system identification among others.

I. A. Arriaga-Trejo (✉)
CONACYT—Autonomous University of Zacatecas, Centro Histórico Zacatecas, Jardín Juárez 147, C. P. 98000 Zacatecas, Mexico
e-mail: iaarriagatr@conacyt.mx

A. G. Orozco-Lugo
Advanced Research Center of the National Polytechnic Institute (CINVESTAV—IPN), Av. Instituto Politécnico Nacional 2508, C. P. 07000 Col. San Pedro Zacatenco, Mexico
e-mail: aorozco@cinvestav.mx

© Springer International Publishing AG, part of Springer Nature 2018   259
M. A. Sanchez et al. (eds.), *Computer Science and Engineering—Theory and Applications*, Studies in Systems, Decision and Control 143, https://doi.org/10.1007/978-3-319-74060-7_14

Recently, the construction of such sequences using numerical techniques has gained considerable attention in the research community. This motivation stems from the applications that sequences with constant magnitude in the time domain (also referred as unimodular) have in reducing interference effects in communications systems and processing of radar signals [5–7, 16–20, 23–27]. In [6], for instance, cyclic algorithms are introduced to generate constant modulus sequences with good aperiodic autocorrelation properties.[1] The main criterion employed in [6] to generate such sequences, is the minimization of the Integrated Sidelobe Level (ISL), which is a popular metric employed to quantify the contribution of the out of phase autocorrelation coefficients. However, the non-convex nature of the cost function makes difficult to locate the points where it vanishes, therefore an *equivalent* target function is proposed, whose global minima are obtained by iteratively using the Fast Fourier Transform (FFT) in the time and frequency domains.

Even though there exists a considerable number of techniques reported to generate unimodular sequences in an efficient manner [6, 7, 23–27], the reader should realize that all of the aforementioned methods base their design criterion on minimizing the contribution of the aperiodic or periodic autocorrelation functions for the out of phase coefficients. This has been the case since the designed sequences are employed to perform strictly linear (SL) processing, where it is only required to process the autocorrelation function. However, it has been shown that improved results can be obtained if the full second order characteristics of processes are considered, by performing widely linear (WL) signal processing [8, 9].

In WL systems, there is an additional degree of freedom that can be exploited when compared to SL systems, which is the complex conjugate of the input signal. Modeling of systems using WL structures has been documented in the literature. In wireless communications systems, WL models arise naturally when non-linear radio frequency (RF) impairments such as in-phase and quadrature-phase (I/Q) imbalances are considered in the analysis of signal propagation [10–12].

The design of *optimal* unimodular sequences for the identification of WL systems has not received that much attention as that for SL systems. Perhaps, the main application of sequences that are adequate to handle WL systems comes from the analyses performed to mitigate interference due to I/Q imbalances in direct conversion transceivers employing multi-carrier schemes such as Orthogonal Division Frequency Modulation (OFDM). For instance, in [13, 14] training sequences are proposed to jointly estimate the channel impulse response of a wireless communication system and interference originated by I/Q imbalances. Nonetheless, no analysis was performed to propose sequences with constant modulus in the time domain.

---

[1]In the specialized literature, by *good* autocorrelation properties it is meant that the out of phase correlation coefficients have considerable low magnitude, ideally zero.

To the best of the authors' knowledge, it was in [16] where sequences with constant magnitude in the time and frequency domains were first reported for the identification of WL systems. The sequences documented, possess an impulse like periodic autocorrelation function with additional restrictions imposed on their periodic complementary autocorrelation function. These features distinctively differentiate them from unimodular sequences for SL system identification.

It is the main objective of this contribution, to instruct the readers in the design of unimodular sequences that are *optimal* for the identification of WL systems. Throughout the manuscript, it will be verified that sequences for WL system estimation are required not only to have good autocorrelation properties, but also good complementary autocorrelation ones. Furthermore, it will be shown that for certain scenarios it is possible to obtain closed form expression for sequences with the desired characteristics, nonetheless in the majority of the remaining cases, numerical techniques are required to generate them.

The material here presented is based upon the contributions made by the authors in the area of WL system identification as well as in the area of sequence design [15–20].

2 Fundamental Concepts

The purpose of this section is to introduce the definitions and terms that will be used through the rest of the chapter in order to develop the theoretical framework of sequences with low complementary autocorrelation function. Here, we also establish the mathematical notation used along the subsequent sections.

Through the rest of this document, we will consider complex sequences of length N, which will be represented by $\{x(n)\}_{n=0}^{N-1}$. The aperiodic autocorrelation function of $x(n)$, which is denoted by $r_{xx}(l)$, is defined to be [5],

$$r_{xx}(l) = \sum_{n=l}^{N-1} x(n)x^*(n-l) \tag{1}$$

for $l = -(N-1), -(N-2), \ldots, N-2, N-1$. We shall refer to each of the values of $r_{xx}(l)$ as the correlation coefficients of the sequence $x(n)$. From the above definition we have the following result.

Lemma 1 *Let $\{x(n)\}_{n=0}^{N-1}$ be a complex sequence of length N. Then the autocorrelation coefficients satisfy, $r_{xx}(l) = r_{xx}^*(-l)$ for $l = -(N-1), -(N-2), \ldots, N-2, N-1$.*

Proof From the definition of the autocorrelation coefficients given by (1), it is verified that,

$$
\begin{aligned}
r_{xx}(-l) &= \sum_{n=-l}^{N-1} x(n) x^*(n+l) \\
&= \sum_{n=-l}^{l-1} x(n) x^*(n+l) + \sum_{n=l}^{N-1} x(n-l) x^*(n) \\
&= \sum_{n=l}^{N-1} x(n-l) x^*(n) \\
&= \left(\sum_{n=l}^{N-1} x(n) x^*(n-l) \right)^* \\
&= r_{xx}^*(-l)
\end{aligned}
\tag{2}
$$

which holds for every $l \in \{-(N-1), -(N-2), \ldots, N-2, N-1\}$.

Similarly, the complementary autocorrelation function of the sequence $\{x(n)\}_{n=0}^{N-1}$, which is denoted by $\gamma_{xx}(l)$, is defined by

$$
\gamma_{xx}(l) = \sum_{n=l}^{N-1} x(n) x(n-l)
\tag{3}
$$

for $l \in \{-(N-1), -(N-2), \ldots, N-2, N-1\}$. Each of the complex numbers $\gamma_{xx}(l)$ will be referred to as the complementary autocorrelation coefficients of the sequence $x(n)$. For the complementary correlation coefficients of a sequence $x(n)$ the following result holds.

Lemma 2 *Let $\{x(n)\}_{n=0}^{N-1}$ be a complex sequence of length N. Then the complementary autocorrelation coefficients satisfy, $\gamma_{xx}(-l) = \gamma_{xx}(l)$ for $l = -(N-1), -(N-2), \ldots, N-2, N-1$.*

Proof From the definition of complementary autocorrelation, given by (3), the involved sum can be expressed as,

$$
\begin{aligned}
\gamma_{xx}(-l) &= \sum_{n=-l}^{N-1} x(n) x(n+l) \\
&= \sum_{n=-l}^{-1} x(n) x(n+l) + \sum_{n=0}^{N-l-1} x(n) x(n+l) \\
&= \sum_{n=l}^{N-1} x(n) x(n-l) \\
&= \gamma_{xx}(l)
\end{aligned}
\tag{4}
$$

By the previous result it can be verified that the complementary autocorrelation function of the sequence $x(n)$ is an even function. Following [20], we shall refer to $\{r_{xx}(l), \gamma_{xx}(l)\}_{l=-(N-1)}^{N-1}$ as the second order characterization of the sequence $x(n)$.

Similar definitions hold for the second order characterization of periodic sequences. For a sequence $x(n)$ of length N, the periodic autocorrelation function, which will be denoted by $\bar{r}_{xx}(l)$, is defined to be,

$$\bar{r}_{xx}(l) = \sum_{n=0}^{N-1} x(n)x^*((n-k) \bmod N) \tag{5}$$

for $l = -(N-1), -(N-2), \ldots, N-2, N-1$. Equivalently, the periodic complementary autocorrelation function, denoted by $\bar{\gamma}_{xx}(l)$ is given by,

$$\bar{\gamma}_{xx}(l) = \sum_{n=0}^{N-1} x(n)x((n-k) \bmod N) \tag{6}$$

for $l = -(N-1), -(N-2), \ldots, N-2, N-1$.

It is quite common to find in the literature metrics used to evaluate the autocorrelation function. The Integrated Sidelobe Level (ISL), for instance, is a common criterion used in the design of sequences with good autocorrelation properties. For a sequence $\{x(n)\}_{n=0}^{N-1}$, the ISL is defined as [6],

$$\text{ISL} = \sum_{l=1}^{N-1} |r_{xx}(l)|^2. \tag{7}$$

The ISL basically measures the contribution of the autocorrelation coefficients for all lags, except that at $l = 0$. Another metric of interest is the Peak to Average Power Ratio (PAPR), which is defined as [4],

$$\text{PAPR} = \frac{\max\left\{|x(0)|^2, \lceil x(1) \rceil^2, \ldots, |x(N-1)|^2\right\}}{\frac{1}{N}\sum_{n=0}^{N-1}|x(n)|^2} \tag{8}$$

In communication systems, it is highly desirable to use sequences with minimum PAPR, in order to avoid exciting nonlinear regions of the power amplifier. As the reader can easily verify, unimodular sequences attain the lowest achievable PAPR, namely PAPR = 1.

In general, we are interested in designing complex sequences with a relative large number of elements. For this reason, in some occasions it will be convenient to employ vector analysis to simplify algebraic manipulations on the derived expressions. Hence, we will denote vectors with boldface lower case characters such as $\mathbf{x}, \mathbf{y}, \mathbf{z}$. Unless otherwise stated, vectors are considered to be column vectors, this is, if $\mathbf{x}$ has M components then it holds, $\dim \mathbf{x} = M \times 1$.

Matrices are denoted with uppercase boldface characters, like $\mathbf{P}, \mathbf{Q}$. The Fourier matrix is denoted by $\mathbf{F}_N$, and its components are given by,

$$
\mathbf{F}_N = \begin{bmatrix} e^{-\frac{2\pi i \cdot 0 \cdot 0}{N}} & e^{-\frac{2\pi i \cdot 0 \cdot 1}{N}} & \cdots & e^{-\frac{2\pi i \cdot 0 \cdot (N-2)}{N}} & e^{-\frac{2\pi i \cdot 0 \cdot (N-1)}{N}} \\ e^{-\frac{2\pi i \cdot 1 \cdot 0}{N}} & e^{-\frac{2\pi i \cdot 1 \cdot 1}{N}} & \cdots & e^{-\frac{2\pi i \cdot 1 \cdot (N-2)}{N}} & e^{-\frac{2\pi i \cdot 0 \cdot (N-1)}{N}} \\ \vdots & \vdots & \ddots & \vdots & \vdots \\ e^{-\frac{2\pi i \cdot (N-2) \cdot 0}{N}} & e^{-\frac{2\pi i \cdot (N-2) \cdot 1}{N}} & \cdots & e^{-\frac{2\pi i \cdot (N-2) \cdot (N-2)}{N}} & e^{-\frac{2\pi i \cdot (N-2) \cdot (N-1)}{N}} \\ e^{-\frac{2\pi i \cdot (N-1) \cdot 0}{N}} & e^{-\frac{2\pi i \cdot (N-1) \cdot 1}{N}} & \cdots & e^{-\frac{2\pi i \cdot (N-1) \cdot (N-2)}{N}} & e^{-\frac{2\pi i \cdot (N-1) \cdot (N-1)}{N}} \end{bmatrix} \tag{9}
$$

and satisfies $\dim \mathbf{F}_N = N \times N$.

The superscripts T, $*$ and H when applied to vectors or matrices indicate the transpose, the conjugate (on each element) and the Hermitian transpose respectively.

Regarding operations related to vector calculus, we denote functions of several variables with italic characters such as f, g. If $f : \mathbb{R}^N \to \mathbb{R}$ is a function from the vector space $\mathbb{R}^N$ to the scalar field $\mathbb{R}$, we represent its gradient by $\frac{\partial}{\partial \mathbf{x}} f(\mathbf{x})$, and its components are given by,

$$
\frac{\partial}{\partial \mathbf{x}} f(\mathbf{x}) = \left[\frac{\partial}{\partial x_0} f(\mathbf{x}), \frac{\partial}{\partial x_1} f(\mathbf{x}), \ldots, \frac{\partial}{\partial x_{N-1}} f(\mathbf{x}) \right]^T, \tag{10}
$$

with $\dim \frac{\partial}{\partial \mathbf{x}} f(\mathbf{x}) = N \times 1$.

3 System Identification Using Cyclostationary Statistics

In this section, we motivate the design of constant modulus sequences for the identification of SL and WL systems. First, we consider the SL system estimation problem by employing the cyclostationary statistics observed at the output of the system. Later on, we extend the reasoning to the design of sequences for WL system estimation.

3.1 SL System Identification

Let us consider the configuration depicted in Fig. 1, where a discrete SL system characterized by a finite impulse response filter (FIR) is given. We want to estimate the coefficients of the filter that define the behavior of the SL system, by making use of the samples observed at the output of the system for a known input sequence $c(n)$. The output is perturbed by Additive White Gaussian Noise (AWGN) that satisfies the following restrictions,

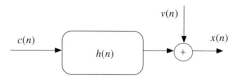

Fig. 1 Strictly linear system with finite impulse response $\{h(n)\}_{n=0}^{M-1}$ to be identified. The system is sounded with a periodic training sequence $c(n)$

$$E\{v(n)\} = 0, \tag{11}$$

and

$$E\{v(n)v^*(m)\} = \sigma_v^2 \delta(m - n). \tag{12}$$

The sequence $c(n)$, used to excite the system is chosen to be periodic with period P, this is, $c(n+P) = c(n)$ for all $n \in \mathbb{Z}$ and mean average power $\sigma_c^2 = P^{-1} \sum_{l=0}^{P-1} |c(l)|^2$. Even though the posed situation is a well-known academic problem [21], there are some subtle aspects that require clarification before proceeding with the analysis.

First of all, the configuration depicted in Fig. 1 is frequently employed to model the *channel estimation* problem in a wireless communication system. In this case, Fig. 1 represents the base band model of a communications system, where $\{h(n)\}_{n=0}^{M-1}$ is a filter with M coefficients resembling the effects of the channel on the transmitted symbols. The main goal for the receiver is to identify the impulse response $h(n)$ from the observed process $x(n)$, assuming knowledge of a training sequence $c(n)$ sent by the transmitter. Once the impulse response $\{h(n)\}_{n=0}^{M-1}$ has been identified, the coefficients of a linear system $\{g(n)\}_{n=0}^{M'}$, with $M' > M$, are computed at the receiver in order to compensate the effects of $h(n)$. The linear system characterized by the impulse response $\{g(n)\}_{n=0}^{M'}$ is known as *equalizer* and it satisfies,

$$(h * g)(n) \approx \delta(n - m_0) \tag{13}$$

for some $m_0 \in \{0, 1, \ldots, M + M' - 2\}$.

Second, when estimating $\{h(n)\}_{n=0}^{M-1}$ using a probing sequence, it is necessary to have an idea of the order of the filter $h(n)$. This is required in order to generate a system of equations that is consistent and enables us to uniquely identify the filter response. Here, without loss of generality, it will be assumed that the period of the training sequence is equal to the number of nonzero coefficients in $h(n)$, this is, $P = M$.

Having stated these facts, we proceed as follows to estimate $h(n)$ from the process $x(n)$ observed at the output of the SL system. The process $x(n)$ is given by,

$$
\begin{aligned}
x(n) &= (h * c)(n) + v(n) \\
&= \sum_{l=0}^{P-1} h(l)c(n-l) + v(n).
\end{aligned}
\tag{14}
$$

From the statistical properties of the noise, and the periodicity of the training sequence $c(n)$, we have that,

$$
E\{x(nP+l)\} = \sum_{m=0}^{P-1} h(m)c(l-m).
\tag{15}
$$

Now, employing the reasoning in [15], if we define $N = N_P P$, with N_P denoting the number of periods used to probe the system, then an estimate of the first order statistics is given by,

$$
\widehat{E}\{x(l)\} = \frac{1}{N_P} \sum_{n=0}^{N_P-1} x(nP+l),
\tag{16}
$$

for $l = 0, 1, \ldots, P-1$. If we define the vector $\widehat{E}\{\mathbf{x}\} \in \mathbb{C}^{P \times 1}$ by

$$
\widehat{E}\{\mathbf{x}\} = \left[\widehat{E}\{x(0)\}, \widehat{E}\{x(1)\}, \ldots, \widehat{E}\{x(P-1)\} \right]^T,
\tag{17}
$$

then it is possible to estimate the impulse response $h(n)$ by using the expression,

$$
\widehat{\mathbf{h}} = \mathbf{C}^{-1} \overline{E}\{\mathbf{x}\}
\tag{18}
$$

where $\mathbf{C}$ is a circulant matrix with elements,

$$
\mathbf{C} =
\begin{bmatrix}
c(0) & c(P-1) & c(P-2) & \cdots & c(2) & c(1) \\
c(1) & c(0) & c(P-1) & \cdots & c(3) & c(2) \\
c(2) & c(1) & c(0) & \cdots & c(4) & c(3) \\
\vdots & \vdots & \vdots & \ddots & \vdots & \vdots \\
c(P-2) & c(P-3) & c(P-4) & \cdots & c(0) & c(P-1) \\
c(P-1) & c(P-2) & c(P-3) & \cdots & c(1) & c(0)
\end{bmatrix},
\tag{19}
$$

such that $\dim \mathbf{C} = P \times P$. The vector $\widehat{\mathbf{h}}$ has dimensions $\dim \widehat{\mathbf{h}} = P \times 1$, and its components are given by $\widehat{\mathbf{h}} = \left[\widehat{h}(0), \widehat{h}(1), \ldots, \widehat{h}(P-1) \right]^T$. Additionally, the estimation error $\mathbf{e}$, is defined to be the difference among the vector containing the channel coefficients and the vector containing their estimates,

$$\mathbf{e} = \mathbf{h} - \widehat{\mathbf{h}}, \tag{20}$$

with $\mathbf{h} = [h(0), h(1), \ldots, h(P-1)]^T$. Besides, the variance of the error is defined as,

$$\sigma_{\mathbf{e}}^2 = \mathrm{tr}\left(E\{\mathbf{e}\mathbf{e}^{\mathbf{H}}\}\right). \tag{21}$$

From the definition of the estimation error, it can be verified that

$$\mathbf{e}\mathbf{e}^{\mathbf{H}} = \frac{1}{N_P^2}\left(\sum_{n=0}^{N_P-1}\sum_{n'=0}^{N_P-1}\mathbf{C}^{-1}\mathbf{v}_n\mathbf{v}_{n'}^{H}(\mathbf{C}^{-1})^{H}\right) \tag{22}$$

with $\mathbf{v}_n = [v(nP), v(nP+1), \ldots, v((n+1)P-1)]^T$. After applying the expectation operator the previous expression reduces to,

$$\begin{aligned}
E\{\mathbf{e}\mathbf{e}^{H}\} &= \frac{\sigma_v^2}{N_P}\mathbf{C}^{-1}(\mathbf{C}^{-1})^{H} \\
&= \frac{\sigma_v^2}{N_P}(\mathbf{C}^{H}\mathbf{C})^{-1}.
\end{aligned} \tag{23}$$

Hence, the variance of the estimation error results from applying the trace operator to (23) which yields,

$$\mathrm{tr}\left(E\{\mathbf{e}\mathbf{e}^{H}\}\right) = \frac{\sigma_v^2}{N_P}\mathrm{tr}\left((\mathbf{C}^{H}\mathbf{C})^{-1}\right). \tag{24}$$

The variance of the estimation error given by expression (23) can be further simplified if another restriction is imposed onto the matrix $\mathbf{C}$, namely that it be unitary, $\mathbf{C}^{H}\mathbf{C} = \mathbf{C}\mathbf{C}^{H} = P\sigma_c^2\mathbf{I}$ as is done in [15]. In order to fulfill this requirement, it is necessary that the columns of $\mathbf{C}$ are orthogonal, this is,

$$(\mathbf{P}^l\mathbf{c})^{H}(\mathbf{P}^m\mathbf{c}) = \begin{cases} P\sigma_c^2 & \text{if } l = m, \\ 0 & \text{if } l \neq m \end{cases} \tag{25}$$

for $l, m \in \{0, 1, \ldots, P-1\}$ with $\mathbf{P}$ being the $P \times P$ permutation matrix with components given by,

$$\mathbf{P} = \begin{bmatrix} 0 & 0 & 0 & \cdots & 0 & 1 \\ 1 & 0 & 0 & \cdots & 0 & 0 \\ \vdots & \vdots & \vdots & \ddots & \vdots & \vdots \\ 0 & 0 & 0 & \cdots & 1 & o \end{bmatrix}, \tag{26}$$

and $\mathbf{c} = [c(0), c(1), \ldots, c(P-1)]^T$.

The condition imposed by expression (25) can be expressed equivalently as,

$$\mathbf{c}^H\left(\mathbf{P}^l\mathbf{c}\right) = \begin{cases} P\sigma_c^2 & \text{if } l = 0 \\ 0 & \text{if } l = 1, 2 \ldots, P - 1. \end{cases} \tag{27}$$

The reader should notice that (27) is the definition of the periodic autocorrelation function of the sequence $\{c(n)\}_{n=0}^{P-1}$. Hence, sequences $\{c(n)\}_{n=0}^{P-1}$, with an impulse-like periodic autocorrelation function, this is, that satisfy (25) or equivalently (27), yield a minimum variance for the estimation error given by,

$$\text{tr}\left(E\{\mathbf{ee}^H\}\right) = \frac{\sigma_v^2}{N_P\sigma_c^2}. \tag{28}$$

As can be observed from (28), the variance of the estimation error decreases as the number of periods used sound the system increases, as expected.

It is important to emphasize that under the assumptions considered for the estimation task, it was found that in order to minimize the variance of the estimation error for SL systems, the selected sequences are required to possess an impulse-like periodic autocorrelation function. In fact, from the analysis developed, no further restriction is required to be imposed on their complementary autocorrelation function.

3.1.1 Second Order Characterization

Sequences with an impulse-like periodic autocorrelation function have been reported in the literature [3, 15]. In fact, it is relatively simple to generate a sequence $\{c(n)\}_{n=0}^{P-1}$ with such properties. As indicated in [22], if we define the vector $\tilde{\mathbf{c}} = \left[e^{i\phi_0}, e^{i\phi_1}, \ldots, e^{i\phi_{P-1}}\right]^T$ with arbitrary values for the phases $\{\phi_m\}_{m=0}^{P-1} \subset [0, 2\pi]$, then the components of the sequence with the desired characteristics are computed using,

$$\mathbf{c} = \mathbf{F}_P^{-1}\tilde{\mathbf{c}} \tag{29}$$

where $\mathbf{F}_P$ is the Fourier matrix with $\dim \mathbf{F}_P = P \times P$. In Fig. 2, the periodic autocorrelation and the periodic complementary autocorrelation functions of 50 sequences are depicted. Each sequence was generated by randomly selecting the phases $\{\phi_m\}_{m=0}^{49}$ from a uniform distribution in the interval $[0, 2\pi]$.

The reader should notice that with the outlined procedure, the generated sequences do not possess constant amplitude in the time domain. However, for practical purposes, besides imposing an impulse-like periodic autocorrelation on sequences, it is also desired that they have constant amplitude in the time domain. In communications systems, for instance, unimodular sequences are desirable in order to avoid exciting nonlinear regions analog components (e.g. power amplifier).

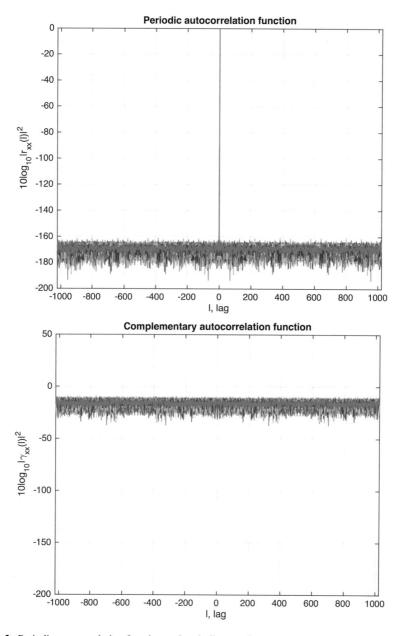

Fig. 2 Periodic autocorrelation function and periodic complementary autocorrelation function for 50 sequences with length $P = 1024$, optimal for SL system identification. The phases of the sequences in the frequency domain were selected randomly from a uniform distribution in the interval $[0, 2\pi]$

Closed form expressions for unimodular sequences with impulse-like periodic autocorrelation function have been documented in the literature, refer for example to [3] and [15]. For a given sequence of length P, a sequence with constant amplitude in the time domain is given by Orozco-Lugo et al. [15],

$$c(n) = \begin{cases} \sigma_c e^{i\frac{\pi}{P}(k(k+1))} & \text{if } P \text{ is odd,} \\ \sigma_c e^{i\frac{\pi}{P}(k(k+2))} & \text{if } P \text{ is even.} \end{cases} \quad (30)$$

It is possible to generate unimodular sequences with good autocorrelation properties as stated in the introduction of the chapter, using numerical methods. Perhaps, the most common technique employed is the Periodic Cyclic Algorithm New (PeCAN) from [7]. Nonetheless, methods based on the majorization-minimization technique have been proposed recently, such as in [23, 26].

3.2 WL System Identification

The purpose of this section is to show that in the estimation of WL systems, it is not only required considering the periodic autocorrelation function, but also the periodic complementary autocorrelation function of the sounding sequence.

Let us consider the WL system depicted in Fig. 3, which is characterized by the impulse responses $\{h_1(n)\}_{n=0}^{M-1}$ and $\{h_2(n)\}_{n=0}^{M-1}$. The output of the system is affected with AWGN, which for simplicity we will further assume that satisfies,

$$E\{v(n)v(m)\} = 0, \quad \forall n, m \quad (31)$$

The system under consideration is excited with a periodic sequence of period P and the output, which is given by,

$$x(n) = (h_1 * c)(n) + (h_2 * c^*)(n) + v(n) \quad (32)$$

is recorded. The problem to solve is to estimate $\{h_1(n)\}_{n=0}^{M-1}$ and $\{h_2(n)\}_{n=0}^{M-1}$ by making use of the information available in the output process $x(n)$ and knowledge of the sounding sequence.

From the stated conditions, it can be verified that there are $2M$ unknowns to be determined[2]; therefore the period of the probing sequence must be selected as $P \geq 2M$. For the sake of simplicity and without loss of generality, we will consider for the remainder of the analysis that $P = 2M$, or equivalently, that each of the filters that define the WL system consists of $P/2$ coefficients. With these restrictions, the output of the system can be written as,

---

[2]It is possible to determine explicitly the number of unknowns in the problem, in this case $\{h_1(n)\}_{n=0}^{M-1}$ and $\{h_2(n)\}_{n=0}^{M-1}$, by expanding the sum in (33).

Fig. 3 Widely linear system with finite impulse responses $\{h_1(n)\}_{n=0}^{M-1}$ and $\{h_2(n)\}_{n=0}^{M-1}$ to be estimated. The system is probed with a periodic training sequence $c(n)$

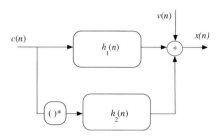

$$x(n) = \sum_{l=0}^{\frac{P}{2}-1} h_1(l)c(n-l) + \sum_{l=0}^{\frac{P}{2}-1} h_2(l)c^*(n-l) + v(n). \tag{33}$$

Following the same reasoning as in the previous section, the first order statistics of the process at the output of the system are given by,

$$E\{x(nP+l)\} = \sum_{m=0}^{\frac{P}{2}-1} h_1(m)c(l-m) + \sum_{m=0}^{\frac{P}{2}-1} h_2(m)c^*(l-m) \tag{34}$$

for $l = 0, 1, \ldots, P-1$. The components of the first order statistics can be arranged in a vector $E\{\mathbf{x}\} = [E\{x(nP)\}, E\{x(nP+1)\}, \ldots, E\{x((n+1)P-1)\}]^T$ satisfying $\dim E\{\mathbf{x}\} = P \times 1$, which relates the training sequence and the impulse responses of the filters in the WL system by,

$$E\{\mathbf{x}\} = \overline{\mathbf{C}}_{\frac{P}{2}}\mathbf{h}_1 + \overline{\mathbf{C}}_{\frac{P}{2}}^*\mathbf{h}_2$$
$$= \overline{\mathbf{C}}\mathbf{h} \tag{35}$$

with $\quad \mathbf{h}_1 = [h_1(0), h_1(1), \ldots, h_1(P/2-1)]^T \quad$ and $\quad \mathbf{h}_2 = [h_2(0), h_2(1), \ldots, h_2(P/2-1)]^T$. The vector $\mathbf{h}$ is in turn defined as $\mathbf{h} = \begin{bmatrix} \mathbf{h}_1^T & \mathbf{h}_2^T \end{bmatrix}^T$ with $\dim \mathbf{h} = P \times 1$. Additionally, the matrix $\overline{\mathbf{C}}_{\frac{P}{2}}$ satisfies $\dim \overline{\mathbf{C}}_{\frac{P}{2}} = P \times \frac{P}{2}$ and has the structure,

$$\overline{\mathbf{C}}_{\frac{P}{2}} = \begin{bmatrix} c(0) & c(P-1) & \cdots & c(\frac{P}{2}+2) & c(\frac{P}{2}+1) \\ c(1) & c(0) & \cdots & c(\frac{P}{2}+3) & c(\frac{P}{2}+2) \\ \vdots & \vdots & \ddots & \vdots & \vdots \\ c(P-2) & c(P-3) & \cdots & c(\frac{P}{2}) & c(\frac{P}{2}-1) \\ c(P-1) & c(P-2) & \cdots & c(\frac{P}{2}+1) & c(\frac{P}{2}) \end{bmatrix}. \tag{36}$$

Meanwhile, $\overline{\mathbf{C}}$ is the augmented matrix constructed as $\overline{\mathbf{C}} = \begin{bmatrix} \overline{\mathbf{C}}_{P/2} & \overline{\mathbf{C}}_{P/2}^* \end{bmatrix}$.

If the arguments employed in the previous section are applied to the WL system estimation problem, then $\widehat{\mathbf{h}}$ is given by,

$$\widehat{\mathbf{h}} = \overline{\mathbf{C}}^{-1}\widehat{E}\{\mathbf{x}\}. \tag{37}$$

The vector $\widehat{E}\{\mathbf{x}\}$ contains the estimates of the first order statistics of the process $x(n)$, this is, $\widehat{E}\{\mathbf{x}\} = \left[\widehat{E}\{x(0)\}, \widehat{E}\{x(0)\}, \ldots, \widehat{E}\{x(P-1)\}\right]^T$. Each of the components of $\widehat{E}\{\mathbf{x}\}$ are computed using,

$$\widehat{E}\{x(l)\} = \frac{1}{N_P}\sum_{n=0}^{N_P-1} x(nP+l), \tag{38}$$

for $l = 0, 1, \ldots, P-1$.

Even though the estimator given by the expression (37) has the same form to that for SL system identification, there is a striking difference among them. Being the structure of the matrix containing the elements of the training sequence. For WL systems, $\overline{\mathbf{C}}$ is an augmented matrix, whereas for the SL systems $\mathbf{C}$ is a circulant matrix.

The nature of the $\overline{\mathbf{C}}$ matrix necessarily imposes restrictions on the training sequence $\{c(n)\}_{n=0}^{P-1}$ that can be employed for the identification of WL systems. As indicated in [16], in order for $\overline{\mathbf{C}}$ to be invertible, $c(n)$ must be complex i.e., it cannot be a purely real sequence. This leads to an important consequence, which is the fact that a WL system cannot be identified using the delta function $\delta(n)$.

Now, from (37) it is possible to derive an expression for the estimation error, which after some algebraic manipulations reduces to,

$$\mathbf{e} = \frac{1}{N_P}\overline{\mathbf{C}}^{-1}\sum_{n=0}^{N_P-1} \mathbf{v}_n \tag{39}$$

with $\mathbf{v}_n = [v(nP), v(nP+1), \ldots, v((n+1)P-1)]^T$.

Furthermore, the variance of the estimation error can be written as,

$$\mathrm{tr}\left(E\{\mathbf{e}\mathbf{e}^H\}\right) = \frac{\sigma_v^2}{N_P}\mathrm{tr}\left(\left(\overline{\mathbf{C}}^H\overline{\mathbf{C}}\right)^{-1}\right). \tag{40}$$

If we impose the restriction that the augmented matrix be unitary, this implies the following set of equations should be simultaneously satisfied,

$$\left(\mathbf{P}^l\mathbf{c}\right)^H\mathbf{P}^m\mathbf{c} = P\sigma_c^2\delta(l-m) \tag{41}$$

and

$$\left(\mathbf{P}^l \mathbf{c}\right)^T \mathbf{P}^m \mathbf{c} = 0 \tag{42}$$

for $l, m \in \{0, 1, \ldots, P/2 - 1\}$. The solution of the system of equations given by (41) and (42) can be easily obtained in the frequency domain. In fact, if we denote by $\widetilde{\mathbf{c}}$ the vector with components $\widetilde{\mathbf{c}} = \left[e^{i\phi_0}, e^{i\phi_1}, \ldots, e^{i\phi_{P-1}}\right]^T$, then the phases $\{\phi_l\}_{l=0}^{P-1}$ must fulfill the following restrictions,

$$\phi_{2l} + \phi_{2\left(\frac{P}{2}-l\right)} = \alpha \tag{43}$$

and

$$\phi_{2l+1} + \phi_{2\left(\frac{P}{2}-l\right)-1} = (\alpha + \pi) \bmod 2\pi \tag{44}$$

with $l = 0, 1 \ldots, 1/2(P/2 - 1)$ for $P/2$ odd, $l = 0, 1, \ldots, P/2$ for $P/2$ even and $\alpha \in [0, 2\pi]$ to be a solution. The elements of the sequence in the time domain are obtained through the relation,

$$\mathbf{c} = \mathbf{F}_P^{-1} \widetilde{\mathbf{c}} \tag{45}$$

as shown in [16].

From the nonlinear system of equations given by (41) and (42), it is possible to notice that the design of sequences for WL system identification involves restrictions on the periodic autocorrelation function (given by (41)) as well as on the periodic complementary autocorrelation function (given by (42)).

The periodic autocorrelation function and the periodic complementary autocorrelation function of 50 sequences of length $P = 1024$ that satisfy the design criterion given by (43) and (44) are depicted in Fig. 4. The phases of the generated sequences were selected randomly from a normal standard distribution. From Fig. 4 it can be readily seen that for WL system identification, the training sequences must possess impulse-like periodic correlation and most important their periodic complementary autocorrelation function should cancel for all lags except those with $l = -P/2$ and $l = P/2$.

As pointed out in [16] and [18], even though it is possible to generate sequences that satisfy the design constraints imposed by (43) and (44), which minimize the variance of the estimation error, a random selection of their phases does not warranty a low PAPR. In [16], it is proposed the use of conventional optimization techniques to generate unimodular sequences that are optimal for the identification of WL systems and satisfy the imposed design restrictions. For that purpose the desired sequence will be selected in such a way that it minimizes the functional $J(\mathbf{\Phi})$,

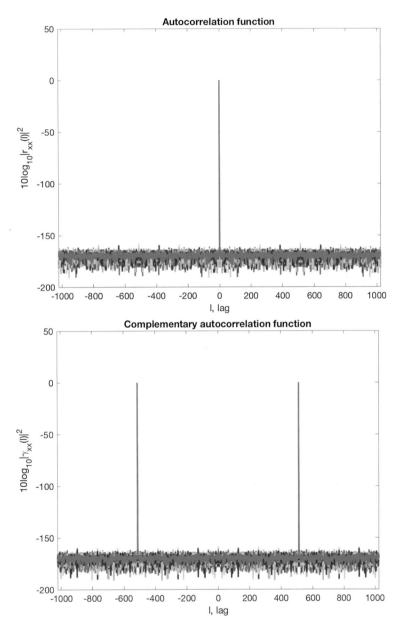

Fig. 4 Periodic autocorrelation and periodic complementary autocorrelation functions of 50 sequences. The phases of the sequences were selected to satisfy (43) and (44)

$$J(\mathbf{\Phi}) = \sum_{n=0}^{P-1} \left(|c(n, \mathbf{\Phi})|^2 - \frac{1}{P} \right)^2 \tag{46}$$

where the notation $c(n, \mathbf{\Phi})$ has been used to explicitly indicate the dependence on the phases $\mathbf{\Phi} = \left[\phi_0, \phi_1, \ldots, \phi_{\frac{P}{2}-1} \right]^T$.

For illustration purposes, Fig. 5 depicts the surface defined by the cost function (46) corresponding to sequences with $P = 6$. As it is shown in [16], due to the symmetries inherent in the design criterion of sequences with constant magnitude in the time domain for the WL scenario, it is possible to reduce the number of unknowns. Furthermore, if we select $\alpha = 0$, from (45) then the coefficients[3] used to generate the surface displayed in the referred figure are given by,

$$c(n, \mathbf{\Phi}) = \frac{i}{3} (-1)^n \sin\left(\phi_1 - \frac{2}{3}\pi n \right)$$
$$+ \cdots + \frac{1}{3} (-1)^n \cos\left(\phi_2 - \frac{1}{3}\pi n \right) \tag{47}$$
$$+ \cdots + \frac{1}{6} (1 + (-1)^n i)$$

where $\mathbf{\Phi} = [\phi_1, \phi_2]^T$ and $n = 0, 1, \ldots, 5$.

With the purpose of exemplifying the performance of sequences with low periodic complementary autocorrelation function for the identification of WL systems, the next numerical simulation is considered. Suppose a WL system having impulse responses $\{h_1(n)\}_{n=0}^{11}$ and $\{h_2(n)\}_{n=0}^{11}$, each of them consisting of 12 coefficients. The WL system is probed with a training sequence $\{c(n)\}_{n=0}^{23}$, whose elements are computed from expression (45). The output of the WL system is affected by AWGN and the process $x(n)$ is used to compute the estimates $\widehat{h}_1(n)$ and $\widehat{h}_2(n)$ using relation (37). The variance of the estimation error obtained with $N_P = 10$, for different values of the average noise power σ_ν^2 is depicted in Fig. 6.

The variance of the estimation error depicted in Fig. 6 was obtained by randomly generating, from a normal standard complex distribution, the impulse responses that define the WL system to be estimated. The estimation error results by averaging numerical values obtained from 100 realizations of the experiment. Each experiment consisted in estimating $\widehat{\mathbf{h}}$, from the process $x(n)$, which is affected by AWGN. The sequence employed in the estimation task, satisfy the design relations given by (43) and (44), with $P = 22$. In order to compare the performance obtained, $\widehat{h}_1(n)$

---

[3]The selection of $\alpha = 0$, necessarily implies $\phi_0 = 0$, as can be verified by direct substitution in (43) and (44).

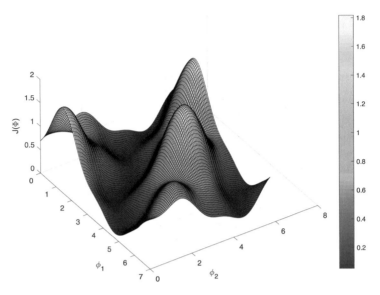

Fig. 5 Surface defined by the cost function given by expression (46) for sequences with period $P = 6$. The minima of the surface define the values of ϕ_1 and ϕ_2 for unimodular sequences that satisfy the imposed constraints for WL system identification

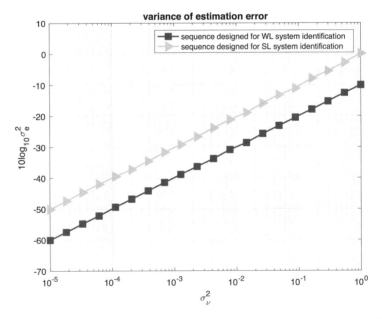

Fig. 6 Variance of estimation error when identifying the impulse responses $\{h_1(n)\}_{n=0}^{11}$ and $\{h_2(n)\}_{n=0}^{11}$ of a WL system. The curve in blue shows the performance obtained when a using sequence that satisfies the restrictions given by relations (41) and (42). Meanwhile, the curve in red shows the performance obtained when employing sequences satisfying the design criterion given by (27). In both cases the period of the sounding sequence was selected as $P = 24$

and $\widehat{h_2}(n)$ were also computed with a sequence possessing an impulse-like periodic autocorrelation function, but with no particular restrictions imposed on its complementary autocorrelation function. From the curves displayed in Fig. 6, it is observed that the variance of the estimation error is considerably lower when using a sequence with good periodic complementary autocorrelation coefficients (as those depicted in Fig. 4).

4 Sequences with Low Aperiodic Complementary Autocorrelation

In the previous section, the design of sequences with specific properties for the periodic autocorrelation function and the periodic complementary autocorrelation function was addressed. From the analysis developed, it was observed that for the estimation of WL systems, the probing sequences must have an impulse-like periodic autocorrelation function and a periodic complementary autocorrelation function that cancels for all lags except for $l = -P/2$ and $l = P/2$. Here, we address the design of sequences whose aperiodic autocorrelation and aperiodic complementary autocorrelation functions cancel for a given set of lags.

4.1 Design of Sequences for an Aperiodic Second Order Characterization

The problem here addressed, is to find the phases of unimodular sequences $\{x(n)\}_{n=0}^{N-1}$ of length N, such that their aperiodic autocorrelation and aperiodic complementary autocorrelation functions cancel for a given set of indexes.

The design of sequences whose aperiodic autocorrelation cancel for a region of interest, has been the subject of research in the literature [6, 23–27]. Among the criteria employed to generate sequences with such properties, the minimization of the ISL is one of the most commonly adopted. In fact, it has been proposed the problem of minimizing a Weighted Integrated Sidelobe Level (WISL), which is defined as [6],

$$\text{WISL} = \sum_{l=1}^{N-1} \alpha_l |r_{xx}(l)|^2, \tag{48}$$

with $\{\alpha_l\}_{l=1}^{N-1} \subset \mathbb{R}$. As has been documented, the minimization of the WISL involves finding the minima of a non-convex function. Therefore, numerical techniques such as iterative methods using FFT [6] or the majorization-minimization technique [23] have been considered in the design of sequences with the desired characteristics.

Now, as can be seen from expression (48), the minimization of the WISL
focuses on imposing restrictions to the autocorrelation function of the sequence
$\{x(n)\}_{n=0}^{N-1}$ to be designed. However, it is not difficult to realize that for WL system
identification, it would be required to consider the complementary autocorrelation
function of the sequence used to probe the system. In [20] the design of unimodular
sequences is carried out by minimizing a generalized WISL, which is defined as,

$$\text{WISL} = \sum_{l=1}^{N-1} \alpha_l |r_{xx}(l)|^2 + \sum_{l=0}^{N-1} \beta_l |\gamma_{xx}(l)|^2 \tag{49}$$

with $\{\alpha_l\}_{l=1}^{N-1} \cup \{\beta_l\}_{l=0}^{N-1} \subset \mathbb{R}$. The reader should notice that in the generalized WISL
given by (49), not only the out of phase coefficients for the aperiodic autocorre-
lation function are penalized, but also those corresponding to the complementary
autocorrelation function.

As it is shown in [20], if we define the vector $\mathbf{x} = \left[e^{i\phi_0}, e^{i\phi_1}, \ldots, e^{i\phi_{N-1}}\right]^T$, with
$\{\phi_l\}_{l=0}^{N-1} \subset [0, 2\pi]$ then the minimization of (49) is equivalent to find the global
minima of the functional,

$$f(\mathbf{\Phi}) = \bar{\mathbf{r}}_{xx}^H(\mathbf{\Phi})\mathbf{M}\mathbf{M}^T\bar{\mathbf{r}}_{xx}(\mathbf{\Phi}) + \bar{\gamma}_{xx}^H(\mathbf{\Phi})\mathbf{M}'\mathbf{M}'^T\bar{\gamma}_{xx}(\mathbf{\Phi}) \tag{50}$$

with $\mathbf{\Phi} = [\phi_0, \phi_1, \ldots, \phi_{N-1}]^T$. The vectors $\bar{\mathbf{r}}_{xx}$ and $\bar{\gamma}_{xx}$ contain the coefficients of the
periodic autocorrelation and the periodic complementary autocorrelation, respec-
tively, of $\bar{\mathbf{x}} = \left[\mathbf{x}^T \mathbf{0}_{N\times1}^T\right]^T$. It should be observed that the vector $\bar{\mathbf{x}}$ contains the
elements of the sequence to be designed appended with N zeros. This has been
done, in order compute efficiently all the correlation and complementary correlation
coefficients of the designed sequence in an efficient manner. Therefore, the vectors
in (50) satisfy $\dim \bar{\mathbf{x}} = \dim \bar{\mathbf{r}}_{xx} = \dim \bar{\gamma}_{xx} = 2N \times 1$. The matrices $\mathbf{M}$ and $\mathbf{M}'$ in turn
have the following structure,

$$\mathbf{M} = [\mathbf{0}_{N\times1}\sqrt{\alpha_1}\mathbf{e}_1 \cdots \sqrt{\alpha_{N-1}}\mathbf{e}_{N\times1}\mathbf{0}_{N\times N}] \tag{51}$$

and

$$\mathbf{M}' = \left[\sqrt{\beta_0}\mathbf{e}_0\sqrt{\beta_1}\mathbf{e}_1 \cdots \sqrt{\beta_{N-1}}\mathbf{e}_{N\times1}\mathbf{0}_{N\times N}\right] \tag{52}$$

where $\{\mathbf{e}_0, \mathbf{e}_1, \ldots, \mathbf{e}_{N-1}\}$ denotes the standard basis for the vector field $\mathbb{C}^{N\times1}$. From
the definition of the matrices $\mathbf{M}$ and $\mathbf{M}'$, the reader should realize that both of them
satisfy $\dim \mathbf{M} = \dim \mathbf{M}' = N \times 2N$.

Due to the nature of the function f to be minimized, which is continuous in each
of the variables ϕ_l, it is possible to find a closed form expression for its gradient and
employ conventional optimization techniques to determine the global minima. The
reader can verify that the gradient is computed from [20],

$$\frac{\partial}{\partial \mathbf{\Phi}} f(\mathbf{\Phi}) = 2\mathrm{Re}\left\{\left(\frac{\eth}{\partial \mathbf{\Phi}} \bar{\mathbf{r}}_{xx}(\mathbf{\Phi})\right)^T \mathbf{M}\mathbf{M}' \bar{r}_{xx}^*(\mathbf{\Phi})\right\}$$
$$+ \cdots + 2\mathrm{Re}\left\{\left(\frac{\partial}{\partial \mathbf{\Phi}} \bar{\gamma}_{xx}(\mathbf{\Phi})\right)^T \mathbf{M}\mathbf{M}'^T \bar{\gamma}_{xx}^*(\mathbf{\Phi})\right\} \tag{53}$$

with

$$\frac{\partial}{\partial \mathbf{\Phi}} \bar{\mathbf{r}}_{xx}(\mathbf{\Phi}) = \left[\frac{\partial}{\partial \phi_0} \bar{\mathbf{r}}_{xx}(\mathbf{\Phi}) \frac{\partial}{\partial \phi_1} \bar{\mathbf{r}}_{xx}(\mathbf{\Phi}) \cdots \frac{\partial}{\partial \phi_{N-1}} \bar{\mathbf{r}}_{xx}(\mathbf{\Phi})\right] \tag{54}$$

and

$$\frac{\partial}{\partial \mathbf{\Phi}} \bar{\gamma}_{xx}(\mathbf{\Phi}) = \left[\frac{\partial}{\partial \phi_0} \bar{\gamma}_{xx}(\mathbf{\Phi}) \frac{\partial}{\partial \phi_1} \bar{\gamma}_{xx}(\mathbf{\Phi}) \cdots \frac{\partial}{\partial \phi_{N-1}} \bar{\gamma}_{xx}(\mathbf{\Phi})\right] \tag{55}$$

both of them satisfying $\dim \frac{\partial}{\partial \mathbf{\Phi}} \bar{\mathbf{r}}_{xx}(\mathbf{\Phi}) = \dim \frac{\partial}{\partial \mathbf{\Phi}} \bar{\gamma}_{xx}(\mathbf{\Phi}) = 2N \times N$.

It should be noticed that for the design of sequences with an aperiodic second order characterization, it is possible to cancel completely the complementary autocorrelation function for a given region of interest.

In order to exemplify the advantages of sequences with low complementary autocorrelation properties, let us consider the identification of a widely linear system with impulse response filters, $\{h_1(n)\}_{n=0}^{L-1}$ and $\{h_2(n)\}_{n=0}^{L-1}$. Furthermore, it will be assumed that the output of the WL system is affected by AWGN. The task to solve is to estimate the coefficients of the filters using a matched filter. This is, the impulse responses that define the WL system, will be estimated using the relations,

$$\widehat{h}_1(n) = \mathbf{x}_n^H \mathbf{y} \tag{56}$$

and

$$\widehat{h}_2(n) = \mathbf{x}_n^T \mathbf{y} \tag{57}$$

with $\mathbf{x}_n = \left[\mathbf{0}_{n \times 1}^T \mathbf{x}^T \mathbf{0}_{(L-n-1) \times 1}^T\right]^T$, where $\mathbf{x}$ contains the sounding sequence and $\mathbf{y}$ contains the samples at the output of the system, satisfying $\dim \mathbf{y} = (N + L - 1) \times 1$.

Figure 7 depicts the variance of the estimation error when a sequence of length $N = 128$, is employed to estimate a WL system with $L = 24$. The probing sequence was obtained by minimizing the cost function given by (50), selecting the weights α_l and β_l such that the autocorrelation and complementary autocorrelation coefficients cancel for $l = 1, 2, \ldots, 23$ and $l = 0, 1, \ldots, 23$ respectively. In order to

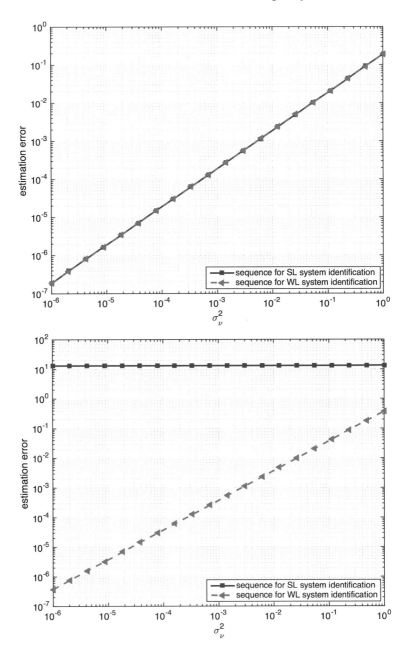

Fig. 7 a Upper: Variance of the estimation error obtained when identifying the impulse response $\{h(n)\}_{n=0}^{23}$ of a SL system. **b** Bottom: Variance of estimation error when identifying the responses $\{h_1(n)\}_{n=0}^{23}$ and $\{h_2(n)\}_{n=0}^{23}$ of a widely linear system

compare the performance obtained with the proposed sequences, the responses $\widehat{h}_1(n)$ and $\widehat{h}_2(n)$ were computed using a sequence with good autocorrelation properties without imposing further restrictions on its complementary autocorrelation function.

5 Conclusions

In this contribution, the authors have addressed the design and applications of sequences with good autocorrelation and good complementary autocorrelation functions. From the analysis developed, it was verified that sequences with particular constraints imposed on their complementary autocorrelation function are better suited to estimate WL systems. The proposed sequences have direct applications for the design of waveforms and for channel estimation in communication systems.

As possible lines of research and future work, the authors identify the design of sets of sequences that have good autocorrelation function and good cross-correlation functions. These sequences can have direct application in the estimation of MIMO systems affected with radio-frequency imperfections such as I/Q imbalances.

Acknowledgements The work of I. A. Arriaga-Trejo is supported by the Mexican Council for Science and Technology (CONACYT) under the Cátedras CONACYT project 3066—Establishment of a Space Telecommunications Laboratory linked to the Mexican Space Agency. The funds are gratefully acknowledged.

References

1. Frank R, Zadoff S, Heimiller R (1962) Phase shift pulse codes with good periodic correlation properties. IRE Trans Inf Theory 8:381–382
2. Schroeder M (1970) Synthesis of low-peak-factor signals and binary sequences with low autocorrelation. IEEE Trans Inf Theory 16:85–89
3. Chu D (1972) Polyphase codes with good periodic correlation properties. IEEE Trans Inf Theory 18:531–532
4. Velazquez-Gutierrez JM, Vargas-Rosales C (2016) Sequence sets in wireless communication systems: a survey. Commun Surveys Tuts 19:1225–1248
5. He H, Li J, Stoica P (2012) Waveform design for active sensing systems—a compu-tational approach. Cambridge Press
6. Stoica P, He H, Li J (2009) New algorithms for designing unimodular sequences with good correlation properties. IEEE Trans Signal Process 57:1415–1425
7. Stoica P, He H, Li J (2009) On designing sequences with impulse-like periodic correlation. IEEE Signal Process Lett 16:703–706
8. Picinbono B, Chevalier P (1995) Widely linear estimation with complex data. IEEE Trans Signal Process 43:2030–2033

9. Adali T, Schreier PJ, Scharf LL (2011) Complex-valued signal processing: the proper way to deal with impropriety. IEEE Trans Signal Process 59:5101–5125
10. Anttila L, Valkama M, Renfors M (2008) Circular based I/Q imbalance compensation in wideband direct-conversion receivers. IEEE Trans Veh Technol 57:2099–2113
11. Schenk TCW, Fledderus ER, Smulders PFM (2007) Performance analysis of zero-IF MIMO OFDM transceivers with IQ imbalance. J Commun 2:9–19
12. Rodriguez-Avila R, Nunez-Vega G, Parra-Michel R et al (2013) Frequency-selective joint Tx/Rx I/Q imbalance estimation using Golay complementary sequences. IEEE Trans Wireless Commun 12:2171–2179
13. Lopez-Estraviz E, De Rore S, Horlin P et al (2007) Pilot design for joint channel and frequency-dependent transmit/receive IQ imbalance estimation and compensation in OFDM-based transceivers. Paper presented at the IEEE international conference on communications, Glasgow, August 2007, pp 4861–4866
14. Minn H, Munoz D (2010) Pilot designs for channel estimation of MIMO OFDM systems with frequency-dependent I/Q imbalances. IEEE Trans Commun 58:2252–2264
15. Orozco-Lugo AG, Lara MM, McLernon DC (2004) Channel estimation using implicit training. IEEE Trans Signal Process 52:240–254
16. Arriaga-Trejo IA, Orozco-Lugo AG, Veloz-Guerrero A et al (2011) Widely linear system estimation using superimposed training. IEEE Trans Signal Process 59:5651–5657
17. Arriaga-Trejo IA, Trancoso J, Vilanueva J et al (2016) Design of unimodular sequences with real periodic correlation and complementary correlation. Electron Lett 52:319–321
18. Arriaga-Trejo IA, Orozco-Lugo AG et al (2016) Joint I/Q imbalances estimation using data-dependent superimposed training. SIViP 11:729–736
19. Arriaga-Trejo IA (2016) Construction of complementary sets of sequences with low aperiodic correlation and complementary correlation. Paper presented at the IEEE global conference on signal and information processing, Washington DC, December 2016, pp 85–89
20. Arriaga-Trejo IA, Orozco-Lugo AG et al (2017) Design of unimodular sequences with good autocorrelation and good complementary autocorrelation properties. IEEE Signal Process Lett 24:1153–1157
21. Scharf LL (1990) Statistical signal processing. Addison-Wesley, Detection, Estimation and Time Series Analysis
22. Frazier MW (1999) An introduction to wavelets through linear algebra. Springer, Berlin
23. Song J, Babu P, Palomar D (2015) Optimization methods for designing sequences with low autocorrelation sidelobes. IEEE Trans Signal Process 63:3998–4009
24. Song J, Babu P, Palomar D (2016) Sequence design to minimize the weighted integrated and peak sidelobe levels. IEEE Trans Signal Process 64:2051–2064
25. Song J, Babu P, Palomar D (2016) Sequence set design with good correlation properties via majorization-minimization. IEEE Trans Signal Process 64:2866–2879
26. Li Y, Vorobyov SA, He Z (2016) Design of multiple unimodular waveforms with low auto- and cross-correlations for radar via majorization–minimization. Paper presented at the 24th european signal processing conference, Budapest, December 2016, pp 2235–2239
27. Zhao L, Song J, Babu P et al (2017) A unified framework for low autocorrelation sequence design via majorization–minimization. IEEE Trans Signal Process 65:438–453

Printed in the United States
By Bookmasters